Modeling and Optimization of Biomass Supply Chains

Modeling and Optimization of Biomass Supply Chains

Top-Down and Bottom-Up Assessment for Agricultural, Forest and Waste Feedstock

Edited by

Calliope Panoutsou
Centre for Environmental Policy, Imperial College London, London, United Kingdom

ELSEVIER

ACADEMIC PRESS
An imprint of Elsevier

Academic Press is an imprint of Elsevier
125 London Wall, London EC2Y 5AS, United Kingdom
525 B Street, Suite 1800, San Diego, CA 92101-4495, United States
50 Hampshire Street, 5th Floor, Cambridge, MA 02139, United States
The Boulevard, Langford Lane, Kidlington, Oxford OX5 1GB, United Kingdom

British Library Cataloguing-in-Publication Data
A catalogue record for this book is available from the British Library

Library of Congress Cataloging-in-Publication Data
A catalog record for this book is available from the Library of Congress

ISBN: 978-0-12-812303-4

For Information on all Academic Press publications
visit our website at https://www.elsevier.com/books-and-journals

Publisher: Joe Hayton
Acquisition Editor: Raquel Zanol
Editorial Project Manager Intern: Gabriela D. Capille
Editorial Project: Mariana Kuhl
Production Project Manager: Sruthi Satheesh
Cover Designer: Greg Harris

Typeset by MPS Limited, Chennai, India

Contents

4. Modeling Biomass Logistics

Bert Annevelink, Perttu Anttila, Kari Väätäinen, Benoît Gabrielle, Daniel García-Galindo, Sylvain Leduc and Igor Staritsky

5. Modeling Demand for Bio-Based Sectors

Marc Londo, Joost van Stralen, Carolien Kraan, Hamid Mozaffarian and Ayla Uslu

List of Contributors

Eija A. Alakangas VTT Technical Research Centre of Finland Ltd, Jyväskylä, Finland

Efthymia Alexopoulou CRES Centre for Renewable Energy Sources and Saving, Pikermi Attikis, Greece

Bert Annevelink Wageningen Food & Biobased Research (WFBR), Wageningen, The Netherlands

Perttu Anttila Natural Resources Institute Finland (Luke), Joensuu, Finland

Ausilio Bauen Centre for Environmental Policy, London, United Kingdom

Pawan Datta University of Freiburg, Freiburg, Germany

Matthias G. Dees Albert-Ludwigs-UniversitätFreiburg Tennebacherstrasse4, Freiburg, Germany; University of Freiburg, Freiburg, Germany

Ilze Dzene WIP Renewable Energies, München, Germany

Berien Elbersen Wageningen Environmental Research, Wageningen, The Netherlands

Wolter Elbersen Wageningen Food & Biobased Research (WFBR), Wageningen, The Netherlands

Joanne Fitzgerald European Forest Institute (EFI), Joensuu, Finland

Nicklas Forsell International Institute for Applied Systems Analysis (IIASA), Laxenburg, Austria

Uwe R. Fritsche IINAS, Darmstadt, Germany

Benoît Gabrielle AgroParisTech/INRA/Paris-Saclay University, Thiverval-Grignon, France

Daniel García-Galindo Research Centre for Energy Resources and Consumption (CIRCE), Zaragoza, Spain

Branko Glavonjic University of Belgrade-Faculty of Forestry, Belgrade, Serbia; University of Belgrade, Belgrade, Serbia

Petar M. Gvero University of Banja Luka, Banja Luka, Bosnia and Herzegovina

Paulien Harmsen Wageningen Food & Biobased Research (WFBR), Wageningen, The Netherlands

Rainer Janssen WIP Renewable Energies, München, Germany

Birger Kerckow Fachagentur Nachwachsende Rohstoffe e.V. (FNR), Gülzow, Germany

Carolien Kraan ECN Energy Research Centre of the Netherlands, Amsterdam, The Netherlands; Stanford University, Stanford, CA, United States

Tijs M. Lammens BTG Biomass Technology Group B.V., Enschedé, The Netherlands

Hans Langeveld Bioenergy Research, Wageningen, The Netherlands

Sylvain Leduc International Institute for Applied Systems Analysis (IIASA), Laxenburg, Austria

Marcus Lindner European Forest Institute (EFI), Joensuu, Finland

Marc Londo ECN Energy Research Centre of the Netherlands, Amsterdam, The Netherlands; Utrecht University, Utrecht, The Netherlands

Andrea Monti University of Bologna, Bologna, Italy

Hamid Mozaffarian ECN Energy Research Centre of the Netherlands, Amsterdam, The Netherlands

Calliope Panoutsou Centre for Environmental Policy, Imperial College London, London, United Kingdom

Christoforos Perakis Centre for Renewable Energy Sources and Saving, Pikermi, Greece

Robert Prinz Natural Resources Institute Finland (Luke), Joensuu, Finland

Jacqueline Ramirez-Almeyda University of Bologna, Bologna, Italy

Raymond Schrijver Wageningen Environmental Research, Wageningen, The Netherlands

Igor Staritsky Wageningen Environmental Research (WER), Wageningen, The Netherlands

Dejan Stojadinovic Energy Transition Consulting, Novi Sad, Serbia

Dragoslava D. Stojiljkovic University of Belgrade, Belgrade, Serbia

Ayla Uslu ECN Energy Research Centre of the Netherlands, Amsterdam, The Netherlands

Kari Väätäinen Natural Resources Institute Finland (Luke), Joensuu, Finland

Joost van Stralen ECN Energy Research Centre of the Netherlands, Amsterdam, The Netherlands

Pieter J. Verkerk European Forest Institute (EFI), Joensuu, Finland

Martijn Vis BTG Biomass Technology Group B.V., Enschede, The Netherlands

Ludger Wenzelides Fachagentur Nachwachsende Rohstoffe e.V. (FNR), Germany

Peter Witzke EUROCARE, Bonn, Germany

Tetiana Zheliezna Institute of Engineering Thermophysics, Kyiv, Ukraine

Sergey Zudin European Forest Institute (EFI), Joensuu, Finland

Acknowledgments

The authors would like to acknowledge that the funding support for the research work presented in this book was primarily based on the three following EU-funded initiatives:

- IEE 08 653 SI2. 529 241. Biomass Futures. Biomass role in achieving the Climate Change & Renewables EU policy targets. Demand and Supply dynamics under the perspective of stakeholders. www.biomassfutures.eu

  The project performed a supply and demand analysis to estimate (through cross energy sector modeling) the role different biomass feedstocks can play in meeting the 2020 RED & NREAP targets across the heat, electricity/CHP, and transport markets.
- IEE/12/835/SI2.645920. Biomass Policies. Strategic Initiative for Resource Efficient Biomass Policies. www.biomasspolicies.eu

  The project built a consistent knowledge base both for the efficient resource mobilization (sustainability criteria; costs, logistics, availability) and for the assessment of resource efficient biomass value chains in EU28 (with a set of consistent technical indicators).
- FP7 KBBE.2013. Grant Agreement n°608622. S2BIOM. Delivery of sustainable supply of non-food biomass to support a "resource-efficient" Bioeconomy in Europe. http://www.s2biom.eu/

The project refined methodologies and provided support for the sustainable delivery of nonfood biomass feedstock at local, regional, and pan European level through developing Strategies, and Roadmaps informed by a "computerized and easy to use" planning toolset (and respective databases) with up-to-date harmonized datasets for EU27, western Balkans, Turkey, Moldova, and Ukraine. The spatial level of analysis both for the toolset and for the databases is NUTS1 (country), NUTS2 (regional), and NUTS3 (local level).

Finally, the editor and the authors would like to thank Mrs. Silvia Vivarelli and Dr. Thomas Schleker, from the European Commission for their continuous guidance and support to the research work presented in this book.

Chapter 1

Biomass Supply Assessments in Europe: Research Context and Methodologies

Calliope Panoutsou, Ausilio Bauen, Berien Elbersen, Matthias G. Dees, Dejan Stojadinovic, Branko Glavonjic, Tetiana Zheliezna, Ludger Wenzelides and Hans Langeveld

1.1 RESEARCH CONTEXT

Since early 2000, several biomass assessment studies were delivered. Most of them were driven by the increasing demand of both policy and industrial actors in the bioenergy and biofuels sectors. As such, the key assumptions used to estimate the available biomass quantities and the respective units in which they have been expressed are strongly related to energy. Their approaches have also been very different; thus, their results are difficult to compare and interpret.

Studies that provide a full-fledged comparison of applied methodologies have been rare till 2010, as they required assessments that integrate multiple sound, well-described, broadly accepted, disciplinary approaches. In USA, the most recent and comprehensive one is the 'Billion ton biomass' study, a national exercise evaluating the potentials and impacts of bioenergy production in the United States. Supported by the Department of Energy and Department of Agriculture, the first study was published in 2005, to be followed by an update in 2011. The second report was authored by 50 scientists from 13 organisations including seven universities. This 235 page report was followed by a new, two-volume, report in 2016 covering over 1000 pages plus numerous background materials including 13 factsheets.

Responding to the same challenges, the European Framework Programme for Research funded the BEE project (http://www.eu-bee.com/) in 2008. The work performed in it provided a state-of-the-art overview for biomass resource assessments in Europe and developed numerous generic approaches, definitions, and a classification of biomass feedstock types to improve the accuracy and comparability of future biomass resource assessments (Rettenmaier et al., 2008; Vis et al., 2010).

Modeling and Optimization of Biomass Supply Chains.
DOI: http://dx.doi.org/10.1016/B978-0-12-812303-4.00001-X

From 2010 onward, research in EU, following similar patterns to the USA, broadened to also analyze biomass uses in value chains for bio-based chemicals, pharmaceuticals, and other materials, and evaluate potential synergies and competition. To accomplish this, both units and respective assumptions have been modified to evaluate the biophysical values of biomass supply potentials at each region, addressing land use, displacement effects, and climate change.

Geographic disaggregation is also a research issue that has been significantly improved during the last 10 years in the analysis of biomass supply potentials. Until 2012, the assessments were made at national level (NUTS0 (https://en.wikipedia.org/wiki/Nomenclature_of_Territorial_Units_for_Statistics)). The Biomass Futures project (www.biomassfutures.eu) used the methodology from the BEE project and provided a detailed Biomass Atlas (http://www.biomassfutures.eu/public_docs/final_deliverables/WP3/D3.3%20%20Atlas%20of%20technical%20and%20economic%20biomass%20potential.pdf) for all European Union (EU) countries at NUTS2 state level (comprises 273 geographic subdivisions of countries in the EU). The Atlas provided spatially detailed and quantified overview of EU biomass potential considering the main criteria determining biomass availability from agricultural and forest feedstocks.

During the period 2013–16, research work build on the respective databases and was further developed by two initiatives:

Biomass Policies (www.biomasspolicies.eu): the work built on the datasets from Biomass Futures in terms of detailed cost supply information and considered competition from known conventional uses (animal feed and food). The following formula has been used for all the feedstocks:

$$Availability = Presence - T_1(-T_2)$$

where:

Availability = Biomass availability given what can be produced, harvested, and collected with current or near future practices and known given state-the-art technologies and taking account of basic environmental sustainability requirements regarding soil and biodiversity conservation.

Presence = Presence of biomass now (and in future given land use change expectations).

T_1 = amount of biomass that should be left behind for soil conservation/biodiversity/erosion control

T_2 = conventional known competitive uses (feed and food)

This work was also performed for all biomass feedstock types (oil, starch, lignocellulosic) with geographic disaggregation at NUTS2 state level.

S2Biom (www.s2biom.eu): the research work built on the BEE, Biomass Futures, and Biomass Policies projects and focused on 50 lignocellulosic biomass types. It covered all Europe, expanded geographic coverage of the biomass supply assessments to non-EU countries in Western Balkans, Moldova, Ukraine, and Turkey, and improved the level of disaggregation to the

TABLE 1.1 Criteria for Selection of Studies for Detailed Analysis

Name	Description
Geographic coverage	To cover all European countries, with focus on studies that provide data at national level
Biomass types	To cover the full range of different biomass types (forestry biomass, nonfood dedicated crops, agricultural residues, and biowastes)
Potential types	To cover the full range of different potential types (theoretical, technical, economic, and implementation potential)
Methods	To cover a representative range of different methods
Time frame	To cover years 2000–30

NUTS3 level in the EU countries, FYR of Macedonia, Serbia, Ukraine, Turkey, which represent 1450 districts across Europe.

1.2 METHODOLOGY AND OVERVIEW OF STUDIES

The methodology used for the comparison of biomass resource assessment studies in this chapter is based on the results of BEE (www.eu-bee.info) and Biomass Futures projects (www.biomassfutures.eu).

- Using the BEE reference list of the main biomass resource assessment studies
- Applying the harmonized definition of biomass assessment methods, biomass types, and biomass potentials.

From the long list of about 250 studies in the BEE/Biomass Futures literature databases, about 40 studies were selected for further analysis that corresponded to the criteria presented in Table 1.1.

To create a consistent overview, the main characteristics of these studies in terms of biomass feedstock types assessed applied methodology and key assumptions are analyzed in terms of terminology, framework conditions, and key assumptions.

1.2.1 Terminology

This section provides an overview of the terminology used by the studies in terms of types of potentials (theoretical, technical, economic, and sustainable) and biomass feedstock categories (Fig. 1.1).

1.2.1.1 Types of Potentials

- Theoretical potential: The theoretical potential is the overall maximum amount of terrestrial biomass which can be considered theoretically available for further exploitation within fundamental biophysical limits.

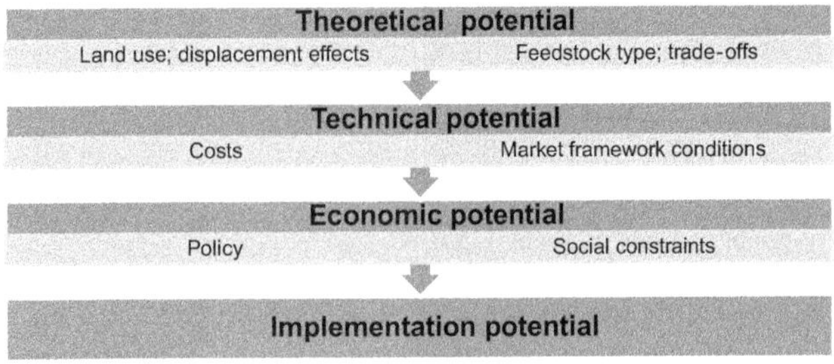

FIGURE 1.1 Definitions of biomass potential types. *Adapted from BEE project, www.eu-bee.com.*

- Technical potential: The technical potential is the fraction of the theoretical potential which is available under given technological options (such as harvesting techniques, infrastructure and accessibility, processing techniques). It also considers spatial restrictions due to other land uses (food, feed, and fiber production) as well as ecological (e.g., nature reserves) and possibly other nontechnical constraints.
- Economic potential: The economic potential is the part of the technical potential which meets criteria of economic rationale within the given framework conditions.
- Implementation potential: The implementation potential is the part of the economic potential that can be implemented within a certain time frame and under concrete socioeconomic and regulatory framework conditions, including economic, institutional and social constraints, and policy incentives.

The classification in types of biomass potentials helps to understand what information is presented. For instance, some biomass types show high technical potentials while their economic potential is rather limited due to the high costs of extraction and transport. Therefore, it is recommended that the type of potential is explicitly mentioned in every biomass resource assessment.

1.2.1.2 Biomass Feedstock Categories

- Agricultural biomass includes grains from oilseed and starch crops (wheat, barley, oats, rye, maize, sunflower, rapeseed, soy), sugar beet, straw, stubbles, woody pruning and orchards residues, grassland cuttings not used for feed purposes, biomass from roadside verges, by-products, and residues from food and fruit processing industry.
- Forest biomass includes (1) primary forestry production from thinnings and final fellings, stem and crown biomass from early thinnings, (2) logging residues and stumps from final fellings, (3) secondary residues from wood industries (sawmill and other wood processing).
- Biowastes include "biodegradable garden and park waste, food and kitchen waste from households, restaurants, catering, and retail premises,

and comparable waste from food processing plants" (Waste Framework Directive, 2008/98/EC).

1.2.2 Framework Conditions and Key Assumptions

This section describes the background of the reviewed studies, regarding time frames, regional coverage, methods, and scenario assumptions used in the estimation. This facilitates better understanding of the type of information provided by these studies.

1.2.2.1 Time Frame

Most of the studies have as baseline years from 2000 to 2005 and include projections for 2010, 2020, and 2030 in response to current policy and support frameworks for renewable, bioenergy, and biofuels.

Only Hoogwijk et al. (2003) assessed the global potential for 17 world regions in 2050–100. Common reference years for the projections in most of the studies are 2010, 2020, and 2020.

1.2.2.2 Geographic Coverage

Most of the reviewed studies provided estimations for EU25 and EU27 while some included Switzerland, Norway, and Ukraine. Two of the studies provided information at global level and the most recent ones cover all Europe, including Western Balkans, Moldova, Ukraine, and Turkey.

1.2.2.3 Approaches

In principle, two approaches have been used to estimate the potentials: (1) bottom-up, or resource based and (2) top-down or demand driven.

- Bottom-up assessments focus mainly on the total biomass resource base and the competition between different uses of the resources (supply side).
- Top-down assessments analyze the competitiveness of biomass-based energy, or estimate the amount of biomass required meeting demand from industry and policy set targets.

Only a few of the most recent studies accounted for costs, environmental impacts, and competitive uses of land from other markets (e.g., food and feed) when estimating the biomass potential of different feedstocks.

1.2.2.4 Data Sources

Most studies use FAO and Eurostat statistics. The spatial analysis is mostly done at NUTS2 and NUTS3 level with CORINE Land Cover being the most commonly used. As the data sources differ (from international and European statistics to national inventories and questionnaire responses), high variability is observed in the results of the studies and thus the possibilities to compare are restricted (Table 1.2).

TABLE 1.2 Framework Conditions for the Reviewed Studies

#	Study	Geographic Coverage	Methods, Framework Conditions, and Key Assumptions	Feedstocks	Time frame
1	EC Biomass Action Plan (2005)	Europe	Resource focused.	All biomass types	2010 (2020, 2030)
2	Schneider et al. (2008)	Europe	Resource focused (land endowments for five land qualities, pasture, and forest), demand and greenhouse gas (GHG) policy driven (each bioenergy option has an exogenously estimated emission coefficient)	Switchgrass, reed canary grass, Miscanthus, poplar, willow, arundo, cardoon, annual crops, trees	2005–max 2050
3	Eriksson and Nilsson (2006)	Europe: EU25 + Belarus + Ukraine	Resource focused. Based on land use area, growth and yields for technical potential; three different assumptions for the three time frames plus two different versions on harvests in forest residues and energy crops (a—low, b—high)	Forest residues, forest industry by-products, crop residues (from trade, agriculture), energy crops	Short term (10–20 years), medium term (20–40 years), long term (>40 years)
4	Krause and Oettel (2007)	Europe: EU27	Resource focused. Biogas potential estimation is based on European databases and survey of previous studies on national and European level. The demand of biogas speeded up in the last years, growing sixfold in the period 1990–2004, with an average yearly increase of 13%	Animal manure, sewage sludge, municipal solid wastes (MSW), industrial wastes, energy crops	2003–05
5	McCormick and Kåberger (2007)	Europe: EU27, case cities in S, AT, FI, UK, I, PL	Methodology to investigate key barriers to expanding bioenergy in the EU, six case studies	Not specified	Current situation
6	Harmelink et al. (2006)	Europe: EU27	Demand driven. In a stepwise approach policy instruments are characterized and analyzed, leading to a quantitative assessment of the likely growth in renewable energy production for each individual technology and country in case no policy changes occur	Not specified	2010

7	Faber et al. (2005)	Europe: EU27; special focus on NL; DE as an illustrative case	Both resource focused and demand driven. Studying the development of relevant policies and legislation, in particular the relevant EC Directives; literature reviews and interviews with experts; reviewing the biomass promoting policies throughout the EU; economic analyses	Biomass for generating electricity and heat. Biomass for chemistry and transport fuels is excluded	Current situation
8	De Noord et al. (2004)	Europe: EU15 plus NO	Resource focused	Energy crops, forestry, solid manure, liquid manure, MSW, agricultural residues, landfill gas, sewage sludge, industrial waste	2000 and sometimes 2030
9	Kloek and Jordan (2005)	Europe: EU15 + 10 + RO + BG + HRr + Turkey + IS + NO + CH	Using an EU "List of Wastes" as described in 2000/532/EC	Organic part MSW, excess manure, construction/demolition wastes	2005–30
10	Siemons et al. (2004)	Europe: EU15 + accession countries plus BG and RO	Modeling of demand, supply, and technology development function. Equilibrium of these function determines the role of biomass as source of renewable energy	Forestry by-products, wood fuels, agricultural residues, industrial residues, energy crops. wet manure, organic waste (biodegradable municipal waste, demolition wood, dry manure, black liquor), sewage gas, landfill gas	2010 and 2020
11	Edwards et al. (2005)	Europe: EU25 + 2	Resource focused. Calculate the straw potential (theoretical), taking the environmental and competitive constrains into account, followed by the economically and logistically availability. Resulting in suitable location for power plants based on the available straw density	Straw from wheat and barley	Actual situation in 2005
12	EEA (2002)	Europe	Resource focused, the starting point of the document is the currently available BMW and the aim is reducing these amounts	Biodegradable municipal waste	1996–2016

(*Continued*)

TABLE 1.2 (Continued)

#	Study	Geographic Coverage	Methods, Framework Conditions, and Key Assumptions	Feedstocks	Time frame
13	Fischer et al. (2007)	Europe	Resource focused. The Agro-Ecological Zones methodology for the potential productivity assessment, estimates cultivated and pasture land availabilities for biofuel feedstock production. Then combined land use and energy yields scenarios	First-generation feedstock groups (cereals, oil crops, sugar crops) and second-generation feedstock groups (herbaceous and woody lignocellulosic plants)	2000–02, 2030
14	Ganko et al. (2008)	Europe	Resource-focused approach. Food and fiber production cannot be affected. Only surplus land available for energy crops	Lignocellulose energy crops (woody and herbaceous), forestry and wood industry by-products, agriculture residues	2020
15	Nikolaou et al. (2003)	Europe	Resource-focused approach. The resource assessment in this study was made in three steps: Technical resource potential, defined as the total annual production of all resources given no limits. Available potential, defined as all resources available with estimated, realistic limits. The energy potential is expressed in PJ/year	Agricultural crop residues, animal wastes, energy crops, forestry residues, wood industry and food processing by-products, biological wastes	2000
16	Junginger et al. (2005)	Europe (case countries Sand NL)	Literature review of (1) multinational biomass comparisons, (2) national biomass policy descriptions and evaluations, and (3) international comparison of general renewable energy policy evaluations	All biomass	Current situation
17	Alakangas et al. (2007)	Europe	Both resource focused and demand driven. By comparing biomass resources and current use, the potential to increase bioenergy production has been estimated	Forest residues, domestic firewood, refined wood fuels, industrial by-products (solid), industrial waste liquors	2001–04, 2010, 2020

18	Asikainen et al. (2008)	Europe	Resource focused. Forest fuels potential includes logging residues from fellings, annual change rate of growing stock and stumps. Annual change rate is calculated as a difference between net annual increment and fellings	Logging residues from fellings, annual change rate of growing stock and stumps	2005
19	EEA (2007)	Europe	Demand driven. The forest biomass potential is modeled using the sustainability classification, projections of future industry demands for wood, international wood trade, and prices of wood. A wide set of sustainable criteria which decrease or increase the forest biomass potential is considered	Felling residues, wood from complementary fellings	2005, 2010, 2020, 2030
20	Hetsch et al. (2008)	Europe	Wood resource balance. Wood flows are analyzed by calculating independently the wood supply and use of wood fibers. The method considers national import and export patterns as well as use and reuse of wood fibers for material and energy purposes. An analysis of historical trends is also included	Industrial wood, fuel wood, logging residues, bark, industrial wood residues, recovered wood, and refined wood fuels	2005, 2010, 2020
21	Meuleman et al. (2005)	Europe	Demand driven. The estimation of biomass potential is based on expansion factors for amounts of forest residues. Projections of wood use for energy are derived from projections of future energy demand for EU	Forest wood biomass	2002, 2010
22	Nielsen et al. (2007)	Europe	Demand driven. Due to the climate changes and constantly increasing emissions it is advised to extend the area covered with energy crops to produce more renewable energy	Energy crops	2030
23	Päivinen et al. (2009)	Europe	Forest area map derived from forest area estimation by Häme et al. (2000), the calibration method by Päivinen et al. (2009), and its implementation by Schuck et al. (2002). Further, regional growing stock statistics referring to regional geographic units were used	Forest wood	1996–2001

(*Continued*)

TABLE 1.2 (Continued)

#	Study	Geographic Coverage	Methods, Framework Conditions, and Key Assumptions	Feedstocks	Time frame
24	Schuck et al. (2002)	Europe	Geometrical correction of the satellite data using polynomial coefficients, separation of the corrected satellite data into three geographic strata, the Atlantic, Mediterranean, and Temperate and Boreal stratum. Estimation of forest variables for each stratum separately. The percentage forest proportion was estimated for each individual pixel of the image mosaic using the CORINE Land Cover classification as training data. The approach of pixel-by-pixel ratio scaling was used for the calibration process and applied to produce a pan-European forest map	Forest wood	1995–2001
25	De Wit et al. (2008)	Europe: EU27 + Switzerland, Norway, Ukraine	Biophysical modeling of land required for food and feed. Agro-Ecological Zones modeling of yields of energy and food crops. Bottom-up cost analysis of bioenergy crops production costs	Annual crops, grasses, and short rotation crops (SRC) poplar/willow on agricultural land, and agricultural and forestry residues	2030
26	EEA (2006)	Europe: EU25	Demand driven There is a need to increase the use of biomass for energy purposes with a strong consideration of environment	Energy crops, forestry biomass, wastes	2030
27	EEA (2012)	EU25	Demand driven.	Agricultural biomass, forestry primary and secondary residues and waste. For agricultural biomass the sub-categories covered are: straw, prunings, manure, dedicated crops on lands not used for food and feed production.	2030

28	Al Saedi et al. (2005)	Europe: specific regions	Resource focused, based on literature review and partner's experience	Biodegradable part of organic wastes and energy crops for biogas production through anaerobic digestion (AD)	2005
29	Al Saedi et al. (2007)	Europe: specific regions	Both resource and demand focused. The work was based on the results of the research carried out in 2002 by a team of Danish researchers. It includes environmental, economic and socio-economic analysis.	Biodegradable part of organic wastes and energy crops for biogas production through AD	2005–07
30	Geletukha et al. (2010)	Ukraine	Technical potential is based on the biomass amount that can be actually collected under existing conditions. Economic potential is determined as a certain share of the technical potential: 33% for cereal crops straw, 100% for rapeseed straw and sunflower residues, 70% for grain maize residues; 90% for felling residues, 50% for wood processing waste	Agricultural and forest biomass	2008
31	Geletukha et al. (2011)	Ukraine	Area under energy crops is determined as currently unused agricultural land (arable land area minus sawn minus fallow land area. For manure, technical potential is based on the biomass amount that can be actually collected under the existing conditions. Economic potential is based on the minimal electrical capacity of a biogas cogeneration unit (200 kW)	Energy crops	2008
32	Synenergy (2011)	Europe: Western Balkans, Moldova, Ukraine	Agricultural residues, forest biomass, energy crops, wastes		
33	Biomass Futures (2012)	Europe: EU27	Both resource and demand focused. The work estimated the role biomass can play to meet the EU targets for 2020, as set in the Renewable Energy Directive	Agricultural residues, forest biomass, energy crops, wastes	2020–30

(*Continued*)

TABLE 1.2 (Continued)

#	Study	Geographic Coverage	Methods, Framework Conditions, and Key Assumptions	Feedstocks	Time frame
34	Ecofys (2013)	Europe: EU27	This study examined a number of waste and residue material and assessed to what extent a "surplus" of the materials exists which can be used to produce biofuels without causing indirect land use changes (ILUC); the rules laid down in the low indirect impact biofuels (LIIB) certification module are used for this purpose	Straw, forestry residues, UCO, corn cobs	2013
35	Biomass Policies (2016)	Europe: EU28		Agricultural residues, forest biomass, energy crops, wastes	2020–30
36	World Bank (2016)	Europe: Western Balkans	Demand driven. The current biomass supply is examined on a two-axis basis: (1) energy use and (2) competing uses. The study also provides a quantification of the main competitive uses of biomass, such as woody biomass in sawmill and panel industry, or straw for animal husbandry. Baseline potential for 2013. The scenarios are based on the following: for agricultural residues—based on harvesting assumptions in local conditions, including EU sustainability criteria; for forest residues—based on official fellings and unregistered consumption (illegal logging), and EU sustainability criteria, for energy crops (miscanthus)—based on FAO estimates on unused agricultural land in Western Balkans countries, and average yield of energy crops (Miscanthus). In addition, production costs for the most promising biomass feedstocks have been examined, as well as the assessment of economically viable biomass options for heating.	Biomass suitable for heating/direct combustion—agricultural residues (straw from cereals and oil crops, corn stover, prunings), forest residues, energy crops (miscanthus)	2013, 2020

37	S2Biom 2016 (S2Biom)	Europe: EU28, Western Balkans, Moldova, Ukraine, Turkey	Supply driven. Lignocellulosic biomass covering 50 feedstock types across Europe	Lignocellulosic biomass from agricultural residues, forest biomass, energy crops, wastes	2012, 2020, 2030
38	Glavonjić et al. (2016)	Europe: EU28, Western Balkans, Moldova, Ukraine, Turkey	The general approach, applied for the residues from sawmills, industries producing semifinished products, and wood-based panels and from pulp and paper industry, is based on the production quantity per industry sector (partly per subsector) per country and the ratio of residues to the production quantity per industry sector/subsector per country.		
			The estimates of totals at NUTS3 level result from a spatial disaggregation of the totals at country level using variables for proportional allocation at NUTS3 level that are assumed to be highly correlated.		

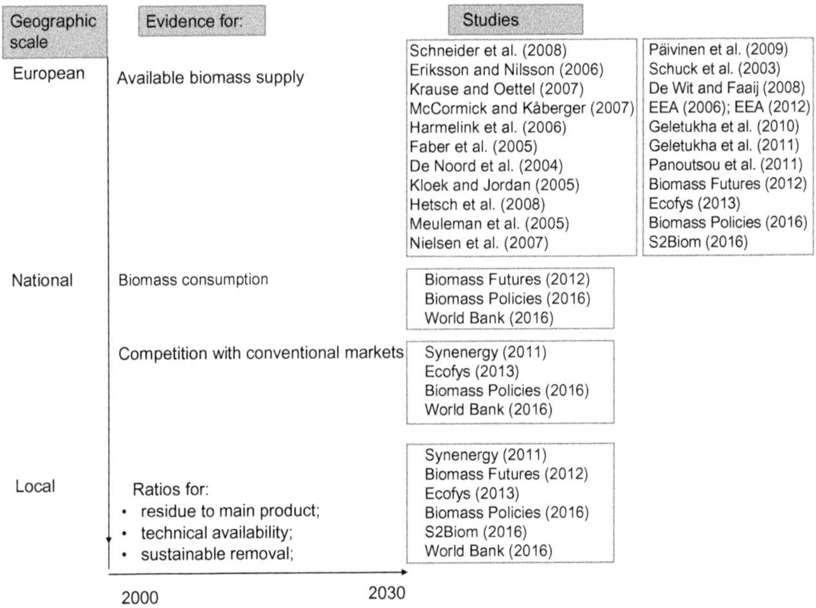

FIGURE 1.2 Evidence provided by current assessments for policy, industry, and research.

1.2.3 Evidence for Policy, Industry, and Research

Fig. 1.2 presents the type of evidence provided by the reviewed studies so far and classifies its relevance for policy, industry, and future research.

1.3 MODELS USED IN BIOMASS SUPPLY ASSESSMENTS

Biomass supply assessments require a matrix of data from different sectoral and regional levels. Fig. 1.3 presents an outline of key data requirements across the whole chains.

There are several modeling approaches that have been used for assessing biophysical biomass availability, land use, sustainability, and impacts. These can be categorized into:

- computable general equilibrium models (van Meijl et al., 2006; Banse et al., 2008; Taheipour and Tyner, 2012; van Meijl et al., 2012; Laborde and Valin, 2012);
- partial equilibrium) models (De La Torre Ugarte and Ray, 2000; Clarke et al., 2009);
- bottom-up models and analyses (Berndes, 2003; Smeets et al., 2007; Trabucco et al., 2010; Muth and Bryden, 2012; de Wit et al., 2008; Marohn and Cadisch, 2011; Cai et al., 2010; van der Hilst et al., 2012; Kurka et al., 2012);
- integrated assessment models (Hoogwijk et al., 2005; van Vuuren et al., 2009; Acosta-Michlik et al., 2011; GEA, 2012; Chaturvedi et al., 2013).

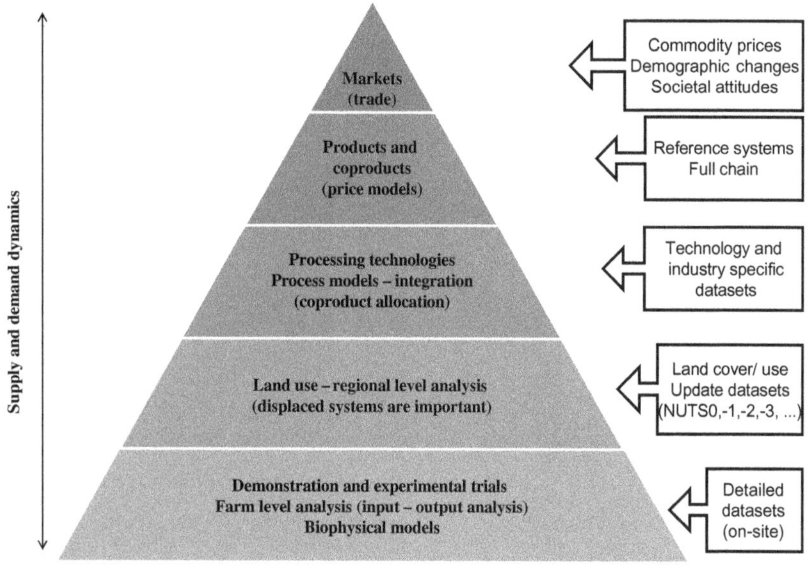

FIGURE 1.3 Data requirements for biomass assessments (Panoutsou, 2008).

This categorization is to some degree artificial because each model tends to have individual characteristics and often includes elements of more than one category (Hourcade et al., 2006).

Wicke et al. (2015) provides a consistent overview of their strengths and weaknesses.

Several European and national projects have developed web-based tools that facilitate integrated spatially explicit availability, suitability, and sustainability performance assessments for bio-based value chains. These have had selected focus, e.g., feedstocks and sustainability issues like the GREET model (Wang et al., 2012) and NitroEurope (2017); rural development and land use change (Volante, 2016); and cost supply of lignocellulosic biomass (S2Biom, 2016). These tools assess availability of biomass feedstocks by location.

1.4 CHALLENGES, GAPS, AND UNCERTAINTIES

1.4.1 Challenges

Based on the overview of the analyzed studies, a set of challenges for biomass supply assessments is described as follows.

1.4.1.1 Definitions for Biomass Potential Types, Categories and Subcategories of Biomass Feedstock Types

Most of the reviewed studies have divided biomass feedstocks per source of origin, i.e., agricultural, forest, and biowaste-derived feedstocks. However, as these categories are broad, several subcategories are defined in the different studies to provide a more detailed analysis for the potentials. This factor is an important difference among the studies and results into different interpretations for the various potentials.

Further to this, the typology of potentials as addressed in Section 1.2.1 is subject to different interpretations and harmonization efforts are required to ensure there is a consistent baseline for estimating the theoretical and technical potentials. The latter represent the biophysical presence and the technical constraints of biomass supply and as such they rely on statistics and indicators from agriculture and forestry that can be refined and applied across countries and regions.

This is not the case for the economic and implementation potentials, though. These are subject to market conditions and prices as well as demand-driven assumptions, and as such they cannot be fully harmonized since they depend largely on local conditions.

1.4.1.2 Data Availability, Homogeneity, and Validity for the Baseline and Land Cover/Use Datasets

The availability, homogeneity, and validity of data rely largely on the official statistics since these are the initial data required for biomass assessments from all sectors (agriculture, forestry, and wastes). In the countries where national statistics are updated on regular basis and appropriate methodologies are developed in the official statistical systems, there are no significant problems and challenges regarding the quality and reliability of the initial data. This is the case for all EU Member States.

However, problems with the reliability of the initial data exist in some of the non-EU countries. Within their official statistical systems, there is no provision for collection, processing, and publishing the data. In these countries, data below the national averages and market prices/costs are particularly scarce as well. In such cases, data collection is restricted and the approach followed so far is based on the detailed validation of the required datasets and indicators by national experts and comparison with local studies.

1.4.1.3 Residue to Main Product and Technical Availability Ratios

The estimation of the theoretical residue yield per hectare depends strongly on the biomass feedstock type and local practices for its management, harvesting, and storage. This is a challenge in biomass supply assessments and requires validation of indicators found in literature from local experts.

Similarly, the technical availability ratios consider technical limitations related to the biomass feedstock physiology, the harvest/collection index, and the prevailing climate and soil conditions in each region. There are several factors that limit the amount of biomass available to be collected and further processed. These include the harvesting equipment, the feedstock type, growth pattern and variety, the harvest index (related to both growth rates and management practices), as well as losses from lodging (e.g., crops flattened by wind or rain), deforestation, forest fires, etc.

1.4.2 Gaps and Uncertainties

Table 1.3 presents important gaps and uncertainties as they emerge from the reviewed studies and links them to the challenges described in the previous section.

1.5 RECOMMENDATIONS FOR FUTURE RESEARCH

Concluding from the findings of the reviewed studies and initiatives, there is a strong need for future research in support of potential studies at national and on European level.

1.5.1 Assessments

The diversity of estimates in the existing assessments is due to several variable factors which can affect scenarios for future demands for biomass and

TABLE 1.3 Gaps and Uncertainties

Area	Gaps and Uncertainties in Biomass Supply Assessments	Challenge
Statistics	Harmonization of terminology and definitions	Definitions
	lack of harmonized European data on land availability, labor, taxation, utilities, crop processing costs, and transportation	Data
	Statistics for biomass require better structure both for supply (agricultural, forest, biowastes) and end use sectors (e.g., energy, bio-based products, and materials)	
	Data on biomass costs and market prices	
Modeling	Integration of issues that are case/site sensitive: • Resource/demand pattern • Conflicts of interests • Logistics	Ratios
Forestry	Productivity of forests and annual harvest index.The supply of forest residues should be linked with the supply of industrial roundwood which is determined by: (1) the demand for wood and paper products, (2) forestry management methods, and (3) soil and climatic factors	

land requirements. These include country/geographic focus; projection times, targets assumed for fossil fuel substitution (either by mass, volume, or anticipated reduction in greenhouse gas (GHG) emissions), type of fossil fuel substitution (solid, liquid, or gaseous substitution), process conversion to bioenergy or biofuels and process efficiency, crops, or feedstocks reviewed; yields, agronomic practices, and land productivity.

Although the existing studies cover a wide range of feedstock types and biomass potentials, they still lack a coherent structure to allow the estimation of potentials which can support policy making. Future efforts should aim for:

- Further development of the existing harmonized methodologies and integration with field research to quantify residue and technical availability ratios at local level.
- Standard setting with criteria and indicators to correspond to residue ratios, technical availability factors, and sustainability criteria will be critical issues for the biomass supply assessments in the immediate future.

1.5.2 Data

Concerning data, there is a clear need for harmonization of datasets, including land cover and land use, statistics on agriculture, forestry and biowastes, as well as data on biomass consumption for energy and nonenergy uses. Better datasets would considerably improve model outputs and the formation of credible assumptions for scenario development. Areas requiring attention include:

- Work toward common datasets including European and international organization/institutes like EEA, JRC, IIASA, OECD, IEA, Eurostat, and FAO. This is expected to increase the credibility of data used by different modeling groups on issues like land area available, types of land use, and trade. Greater cooperation will also reduce costs and duplication.
- Generation of a harmonized inventory of permanent crops (olive, wine, fruit) throughout Europe integrating already existing databases and remote sensing information (for general agricultural issues as well as for energy issues).
- Grasslands to be used for bioenergy production—assessment of potentials, limitations and technical, as well as economic feasibility (cross-cutting from remote sensing—economy—to technical solutions for "green biorefinery")
- Further development of Earth Observation (EO) products tailored to user's needs and modeling requirements.
- Further integration of economic and implementation issues in EO-based spatially explicit potential assessments.
- Development and demonstration of methods for a high integration of using EO data and in situ monitoring systems for bioenergy potential assessments.

- Combined evaluation of biomass as feedstock for all energy and nonenergy sectors aiming to use primary biomass as feedstock.
- Education and training for countries/organization not yet familiar with the use of EO satellite data.
- Assessment of future yield increases in relation to land and water inputs.
- Trends in biomass use for food, feedstock, fiber, and biorefinery, and their impact on biomass potentials for energy.
- Expand environmental analysis beyond GHG balances to include water issues, soils, biodiversity.
- Harmonize assumptions on implementation and governance of biomass policies/sustainability standards.
- Include potential impacts of climate change (determination and verification of CO_2 avoidance).

1.5.3 Collaboration

Future improvements in the field of biomass supply assessments in Europe should take stock of knowledge developed and further:

- Improve collaborations among existing institutions and networks both on the supply (forestry, agriculture, waste) and the end use (energy, bio-based products, and materials) sides.
- Link efforts of the EO scientific community which the ones from biomass experts to take full advantage of the existing European capacities and infrastructures.
- Establish regular links between entities working on national and EU level assessments in a dedicated network. The network should include institutions and experts from the scientific, industrial, and local governments sectors with good understanding of local context.
- Strengthen cross-sectoral cooperation at national level among institutions and organizations dealing with the biomass feedstocks and among institutions and organizations dealing with their end uses, especially in the countries in which the existing level of cross-sectoral cooperation is still very low.

1.6 CONCLUSIONS

During the period 2008–16, significant improvements occurred in the field of biomass supply assessments in Europe in terms of both harmonizing methodologies and their underlying assumptions. Thus, the first studies have been carried out on European level with detailed disaggregation at district level of statistics covering a wide number of biomass feedstock types and their theoretical, technical, economic, and implementation potentials.

This was particularly important for the non-EU countries (Western Balkans, Moldova, Ukraine, and Turkey) in which the methodologies were used for the first time to harmonize the potentials and align their data and scientific evidence with the potentials of the EU Member States.

Future assessments must consider the role of sustainable practices and the development of appropriate infrastructures in agriculture and forestry for both energy and bio-based product and materials applications. The analysis of regional and local framework conditions is critical for the accuracy of the findings. Ecological and socioeconomic conditions especially are different from place to place, and the implementation of biomass production and utilization chains (both for regional use and for European and the world market) is a local issue.

REFERENCES

Acosta-Michlik, L., Lucht, W., Bondeau, A., Beringer, T., 2011. Integrated assessment of sustainability trade-offs and pathways for global bioenergy production: framing a novel hybrid approach. Renew. Sustain. Energ. Rev. 15, 2791–2809.

Asikainen, A., Liiri, H., Peltola, S., Karjalainen, T., Laitila, J., 2008. Forest Energy Potential in Europe (EU 27), 69. Working Papers of the Finnish Forest Research Institute, 33 p.

Alakangas, E., Heikkinen, A., Lensu, T., Vesterinen, P., 2007. Biomass fuel trade in Europe. Summary report VTTR0350807. Jyväskylä, Finland, VTT57.

Banse, M., van Meijl, H., Tabeau, A., Woltjer, G., 2008. Will EU biofuel policies affect global agricultural markets? Eur. Rev. Agric. Econ. 35, 117–141.

BEE project (Biomass Energy Europe). <www.eu-bee.com>.

Berndes, G., Hoogwijk, M., Van den Broek, R., 2003. The contribution of biomass in the future global energy supply: a review of 17 studies. Biomass Bioenergy 25 (1), 1–28.

Biomass Futures, 2012. Final technical report. <www.biomassfutures.eu>.

Biomass Policies, 2016. www.biomasspolicies.eu Biomass supply for a bioenergy and bioproducts industry. R.D. Perlack and B.J. Stokes (Leads), ORNL/TM-2011/224. Oak Ridge National Laboratory, Oak Ridge, TN. 227p.

Cai, X., Zhang, X., Wang, D., 2010. Land availability for biofuel production. Environ. Sci. Technol. 45, 334–339.

Clarke, D., Jablonski, S., Moran, B., Anandarajah, G., Taylor, G., 2009. How can accelerated development of bioenergy contribute to the future UK energy mix? Insights from a MARKAL modelling exercise. Biotechnol. Biofuels 2, 13.

De Noord, M., Beurskens, L., De Vries, H., 2004. Potentials and costs for renewable electricity generation. A data overview. ECN-C-03-006.

Chaturvedi, V., Hejazi, M., Edmonds, J., Clarke, L., Kyle, P., Davies, E., Wise, M., 2013. Climate mitigation policy implications for global irrigation water demand. Mitigation Adapt. Strategies Global Change. Available from: http://dx.doi.org/10.1007/s11027-013-9497-4.

De La Torre Ugarte, D.G., Ray, D.E., 2000. Biomass and bioenergy applications of the POLYSYS modeling framework. Biomass Bioenerg. 18, 291–308.

De Wit, M.P., Faaij, A.P.C., Fischer, G., Prieler, S., Van Velthuizen, H.T., 2008. Biomass Resources Potential and Related Costs. The Cost-Supply Potential of Biomass Resources in the EU-27. Copernicus Institute, Utrecht University, the Netherlands and the International Institute of Applied Systems Analysis (IIASA), Austria, Switzerland, Norway and the Ukraine. Utrecht, the Netherlands and Laxenburg, Austria, p. 122.

Edwards, R.A.H., Šúri, M., Huld, T.A., J.F. Dallemand, 2005. GIS-based assessment of cereal straw energy resource in the European Union.

Ericsson, K., Nilsson, L.J., 2006. Assessment of the potential biomass supply in Europe using a resource-focused approach. Biomass Bioenerg. 30 (1), 1–15.

European Environment Agency, 2002. Review of selected streams: sewage sludge, construction and demolition waste, waste oils, waste from coal fired power plants. Technical report 69.

EC Biomass Action Plan, 2005. <http://eur-lex.europa.eu/legal-content/EN/TXT/?uri = URISERV: l27014> (accessed 31.01.17.).

EEA—European Environmental Agency, 2006. How much bioenergy can Europe produce without harming the environment?.

Ecofys, 2013. Low ILUC potential of wastes and residues for biofuels. Straw, forestry residues, UCO, corn cobs. September 4, 2013. Project number: BIEDE13386 / BIENL12798.

EEA, 2007. Environmentally Compatible Bio-Energy Potential from European Forests. European Environment Agency (EEA), Copenhagen, Denmark, 39 + Appendices.

ETC/SIA, 2013. Review of the EU Bioenergy Potential From a Resource Efficiency Perspective. Background Report to EEA Study. Alterra, Wageningen.

Hetsch, S., 2008. European wood potential- enough to satisfy future energy and industry needs? 9thconference of the European state forestry organizations. 9–12 June, Bialowieza, Poland. European biomass harvesting technologies (PDF Download Available). Available from: <https://www.researchgate.net/publication/236886591_European_biomass_harvesting_technologies> (accessed 20.04.17.).

Hoogwijk, M., Faaij, A., Van den Broek, R., Berndes, G., Gielen, D., Turkenburg, W., 2003. Exploration of the ranges of the global potential of biomass for energy. Biomass Bioenergy 25, 119–133.

van der Hilst, F., Verstegen, J.A., Karssenberg, D., Faaij, A.P.C., 2012. Spatiotemporal land use modelling to assess land availability for energy crops—illustrated for Mozambique. Global Change Biol. Bioenerg. 4, 859–874.

Fischer, G., Hizsnyik, E., Prieler, S., Van Velthuizen, H., 2007. Assessment of Biomass Potentials for Bio-Fuel Feedstock Production in Europe: Methodology and Results. International Institute for Applied Systems Analysis (IIASA), Laxenburg, Austria, p. 81.

Faber, J., Bergsma, G., Vroonhof, J., 2005. Bioenergy in Europe. Policy Trends and Issues. CE, Delft, The Netherlands, p. 64.

Hoogwijk, et al., 2005. Potential of biomass energy out to 21000, for four IPCC SRES land-use scenarios.

Ganko E., Kunikowski G., Pisarek M., Rutkowska-Filipczak M., Gumeniuk A., Wróbel A., 2008. Biomass resources and potential assessment. Final report WP 5.1.

Glavonjić et al., 2016. Secondary forest residues assessments. S2Biom project. <http://www.s2biom.eu/en/publications-reports/s2biom.html>.

Geletukha, G.G., Kucheruk, P.P., Matveev, Yu.B., Khodakivska, T.V., 2011. Prospects for biogas production in Ukraine. Renew. Energ. N3, 73–77 (in Ukrainian).

Geletukha, G.G., Zheliezna, T.A., Zhovmir, M.M., Matveev, Yu.B., Drozdova, O.I., 2010. Assessment of biomass energy potential in Ukraine. Part 1. Agricultural residues and wood biomass // Industrial Heat Engineering v. 32 (N 6), 58–65.

Junginger, M., Faaij, A., Björheden, R., Turkenburg, W.C., 2005. Technological learning and cost reductions in wood fuel supply chains in Sweden. Biomass Bioenergy 29 (6), 399–418.

Krause, H., Oettel, E., 2007. Key barriers for bioenergy in Europe: economic conditions, know how, institutional capacity and supply chain coordination. REDUBAR Technical Progress Report. DBI, Gas and Umwelttechnik, Leipzig. Germany, p. 129.

Hourcade, J.C., Jaccard, M., Bataille, C., Ghersi, F., 2006.). Hybrid modeling: new answers to old challenges. Energy J. 2, 1–12.

Kurka, T., Jefferies, C., Blackwood, D., 2012. GIS-based location suitability of decentralized, medium scale bioenergy developments to estimate transport CO_2 emissions and costs. Biomass Bioenerg. 46, 366–379.

Laborde, D., Valin, H., 2012.). Modeling land-use changes in a global CGE: assessing the EU biofuel mandates with the Mirage-BioF model. Climate Change Econ. 3, 1250017.

Kloek, W., Jordan, K., (Eds.), 2005. Waste Generated and Treated in Europe–Data 1995–2003.

McCormick, K., Kåberger, T., 2007. Key barriers for bioenergy in Europe: economic conditions, know-how and institutional capacity, and supply chain coordination. Biomass Bioenergy 31 (7), 443–452.

Marohn, C., Cadisch, G., 2011. Documentation and Manual of the LUCIA Model Version 1.2, Institute of Plant Production and Agroecology in the Tropics and Subtropics. University Hohenheim, Stuttgart, Germany.

Meuleman, B., Kuiper, L., Nabuurs, G., 2005. EFFECT. EU Forests for Renewable Energy to Mitigate Climate. European Commission, Brussels, Belgium.

Muth Jr., D.J., Bryden, K.M., 2012. An integrated model for assessment of sustainable agricultural residue removal limits for bioenergy systems. Environ. Modell. Softw. 39, 50–69.

NitroEurope, 2017. <http://www.nitroeurope.eu/> (accessed 31.01.17.).

Nielsen, J.B.H., Oleskowicz- Popiel, B., Al Seadi, 2007. Energy Crops Potentials for EU27. Berlin-Germany. 15th European Biomass Conference and Exhibition.

Päivinen, R., Van Brusselen, J., et al., 2009. The growing stock of European forests using remote sensing and forest inventory data. Forestry 82 (5), 479–490.

Panoutsou, C., 2008. "Biomass & Energy Production" 2008. "X ESA Congress": Agriculture as Resource for Energy and Environmental Presentation—Multifunctional Agriculture, Bologna.

Rettenmaier, N., Reinhardt, G., Schorb, A., Köppen, S., Von Falkenstein, E., 2008. and other BEE partners. Status of Biomass Resource Assessments, BEE Project Deliverable 3.6.

Schuck, A., Van Brusselen, J., Päivinen, R., Häme, T., Kennedy, P., Folving, S., 2002. Compilation of a Calibrated European Forest Map Derived from NOAA-AVHRR Data. Report 13. EFI Internal. European Forest Institute, Joensuu.

Siemons, R., Vis, M., v.d. Berg, D., Chesney, I.M., Whiteley, M., Panoutsou, C., Nikolaou, N., 2004. Bioenergy's role in the EU energy market—a view of developments until 2020.

Schneider, U.A., Balkovic, J., De Cara, S., Franklin, O., Fritz, S., Havlik, P., et al., 2008. The European Forest and Agricultural Sector Optimization Model (EUFASOM), Paper prepared for the 16th annual Conference of the European Association of Environmental and Resource Economists (EAERE), 47. Hamburg University, Hamburg, Germany.

Smeets, E.M.W., Faaij, A.P.C., Lewandowski, I.M., Turkenburg, W.C., 2007. A bottom-up assessment and review of global bio-energy potentials to 2050. Progr. Energ. Combust. Sci. 33 (1), 56–106.

S2Biom project, 2016. <www.s2biom.eu>.

Synenergy, 2011. USAID—HELLENIC AID Energy Cooperation. Project Code: 93.41.028.31. Action 1: Regional Renewable Energy Assessment. Biomass in the Western Balkans, Moldova & Ukraine. Report No. 1.4-26.EC.10.F.

Taheipour F., Tyner W., 2012. Welfare assessment of the renewable fuel standard: economic efficiency, rebound effect and policy interaction in a general equilibrium framework. 15th Annual Conference on Global Economic Analysis. Geneva.

Trabucco, A., Achten, W.M.J., Bowe, C., Aerts, R.A.F., Orshoven, J.V., Norgrove, L., et al., 2010. Global mapping of Jatropha curcas yield based on response of fitness to present and future climate. Global Change Biol. Bioenerg. 2, 139–151.

van Meijl, H., van Rheenen, T., Tabeau, A., Eickhout, B., 2006. The impact of different policy environments on agricultural land use in Europe. Agriculture, Ecosystems & Environment 114, 21–38.

van Meijl, H., Smeets, E.M.W., van Dijk, M., Powell, J., 2012. Macro-Economic Impact Study for Bio-based Malaysia. LEI report 2012-042. WUR-LEI. The Hague.

Vis, M., Berg, D., van den et al., 2010. BEE-D5.1: Harmonisation of biomass resource assessments. Vol. I. Best practices and methods handbook. FP7 grant agreement no. 213417.

Volante, 2016. <http://www.eficeec.efi.int/portal/projects/volante/> (accessed 31.01.17).

van Vuuren, D.P., van Vliet, J., Stehfest, E., 2009. Future bio-energy potential under various natural constraints. Energ. Policy 37, 4220–4230.

Waste Framework Directive, 2008. Directive 2008/98/EC on waste (Waste Framework Directive). <http://ec.europa.eu/environment/waste/framework/>.

Wang, M., Han, J., Dunn, J.B., Cai, H., Elgowainy, A., 2012. Well-to-wheels energy use and greenhouse gas emissions of ethanol from corn, sugarcane and cellulosic biomass for US use. Environ. Res. Lett. 7, 045905.

Wicke, B., Van der Hilst, F., Daioglou, V., Banse, M., Beringer, T., Gerssen Gondelach, S., et al., 2015. Model collaboration for the improved assessment of biomass supply, demand, and impacts. GCB Bioenergy 7, 422–437. Available from: http://dx.doi.org/10.1111/gcbb.12176.

World Bank, 2016. Glavonjić B., Perakis, Ch. and D. Stojadinovic. Task 1: Biomass supply potentials for heat generation in Western Balkans.

FURTHER READING

Allen, J., Browne, M., Hunter, A., Boyd, J., Palmer, H., 1996. Logistics management and costs of biomass fuel supply. Int. J. Phys. Distrib. Logist. Manage. 28 (6), 463–477.

Dees, M., Elbersen, B., Fitzgerald, J., Vis, M., Anttila, P., Forsell, N., et al., 2016. A spatial data base on sustainable biomass cost–supply of lignocellulosic biomass in Europe—methods & data sources. Project report. S2BIOM—a project funded under the European Union 7th Frame Programme. Grant Agreement no. 608622. University of Freiburg, 170 pp.

Fischer, et al., 2005. Biomass potentials of miscanthus, willow and poplar: results and policy implications for Eastern Europe, Northern and Central Asia.

Glavonjić, B., Vukadinović, M., 2014. Efficiency of firewood utilization in households in the pilot regions in Serbia, E4tech (UK) Ltd for GIZ DKTI Programme.

Glavonjić, et al., 2014. Woodfuel Integrated Supply/Demand Overview Mapping (WISDOM) for Montenegro, FAO and Luxembourg Agency for Development Cooperation.

Glavonjić, B., et al., 2015. Spatial wood fuels production and consumption analysis for Serbia, FAO.

Kim, S., Dale, B.E., 2004. Global potential bioethanol production from wasted crops and crop residues. Biomass Bioenerg. 26, 361–375.

Langholtz, M.H., Stokes, B.J., Eaton, L.M., 2016. Advancing Domestic Resources for a Thriving Bioeconomy. Volume 1: Economic Availability of Feedstocks. (Leads), ORNL/TM-2016/160. Oak Ridge National Laboratory, Oak Ridge, TN, 448p. http://dx.doi.org/10.2172/1271651 http://energy.gov/eere/bioenergy/2016-billion-ton-report.

Nikolaoy, A., Lychnaras, V., Panoutsou, C., 2002: Characteristics and geographical distribution of agricultural residues for energy production in Greece. 12th European Conference and Technology Exhibition Biomass for Energy and Industry. Amsterdam. June 2002.

Obersteiner, M., et.al., 2006. Global supply of biomass for energy and carbon sequestration from afforestation/reforestation activities.

Panoutsou, C., Eleftheriades, J., Nikolaou, A., 2009. Biomass supply in EU27 from 2010 to 2030. Energ. Policy 37 (2009), 5675–5686. Available from: http://dx.doi.org/10.1016/j.enpol.2009.08.032.

REFUEL, 2008. Socio-economic and barrier assessment. WP6. Final report. <http://www.refuel.eu/fileadmin/refuel/user/docs/Refuel_D15__WP6__report.pdf>.

Stojadinovic, D., De Bortoli, E., Baldini, M., 2017. Sector Study on Biomass-Based Heating in the Western Balkans. World Bank.

Volk, T., Erjavec, E., Mortensen, K., 2014. Agricultural Policy and European Integration in South-eastern Europe. FAO, Budapest.

Geletukha, G.G., Zheliezna, T.A., Zhovmir, M.M., Matveev, Yu.B., Drozdova, O.I., 2011. Assessment of biomass energy potential in Ukraine. Part 2. Energy crops, liquid biofuels, biogas. Indus. Heat Eng. 33 (1), 57–64.

Chapter 2

Existing Modeling Platforms for Biomass Supply in Europe

Berien Elbersen, Nicklas Forsell, Sylvain Leduc, Igor Staritsky, Peter Witzke and Jacqueline Ramirez-Almeyda

2.1 INTRODUCTION

Biomass availability is related to land use patterns as these deliver different types of biomass, e.g., stem wood, forest residues, dedicated crops, crop residues, and grassland cuttings.

Modeling frameworks are in place to make predictions of future land use changes taking into account certain demands for land to produce crops, livestock, timber and feedstock for bioenergy, biochemicals, and biomaterials. To increase the validity of model findings, these assessments should account for environmental, economic, and socioeconomic constraints such as soil organic matter, maintenance of tropical rain forests and untouched boreal forests, ecosystem services, income, population growth, and food-diet shifts. Additional constraints are the increased biomass demand for energy and bio-based products which is strongly steered by policies and societal demands to reduce climate change impacts, e.g., GHG emissions and respective dependence on fossil feedstocks.

Given the variety of factors influencing land use changes now and in the future, it is quite challenging and complex to predict future patterns and related biomass production for different demands. This process requires an integration of several sectors such as the agricultural, forest, energy, and bio-based products in one model or in a chain of models. This chapter explains which models are currently used in Europe to predict future land use changes. Special attention is paid to the CAPRI and GLOBIOM that have been used in a series of assessments for the last 15 years. Both models are able to assess domestic biomass cost supply potentials in Europe and the biomass cost-supply potential for imports from the rest of the world (RoW). Recent work in S2BIOM and Biomass Policies projects, using the models is described. The chapter concludes with a discussion of the advantages and limitations of such models and recommendations for future work in the field.

Modeling and Optimization of Biomass Supply Chains.
DOI: http://dx.doi.org/10.1016/B978-0-12-812303-4.00002-1

2.2 CURRENT DEVELOPMENTS IN LAND USE MODELING APPROACHES FOR ASSESSING BIOMASS AVAILABILITY IN EUROPE

Agricultural and forest land use patterns are strongly influenced by the respective market demands. These consist of demand for food, feed and in the last decades also increasingly for non-food uses, particularly bioenergy and biobased products. This latter non-food related demand is strongly driven by policies such as the Renewable Energy Directive (RED) (2009/EC/28) and the Fuel Quality Directives (Directive (EU) 2015/1513 (Directive (EU) 2015/1513 of the European Parliament and of the Council of September 9, 2015, amending Directive 98/70/EC relating to the quality of petrol and diesel fuels and amending Directive 2009/28/EC on the promotion of the use of energy from renewable sources) all aiming at decarbonization of the economy. (The European Commission adopted a strategy to shift the European Economy toward greater and more sustainable use of renewable biological resources—"Innovating for Sustainable Growth—a Bioeconomy for Europe" in 2012 (see http://ec.europa.eu/research/bioeconomy/pdf/201202_innovating_sustainable_growth_en.pdf.) Increased biomass feedstock use both in energy and biobased industries is seen as an important instrument to bring GHG emissions down and thus decarbonize the economic sectors still relying most strongly on fossil feedstock.

Similarly, outside Europe there are many other countries, including Canada, China, India, South Africa, and the United States, that have set ambitious targets for increasing the use of bioenergy, especially for liquid biofuels in the transport sector, also putting additional pressure on land.

Land demand is a sensitive issue as it is a scarce resource and many ecosystem services are related to good management of it. Because of this, robust estimates of land availability for biomass production for nonfood uses cannot be projected without taking the complex sustainability constraints into account.

Modeling results are first giving the level of total volumes of agricultural and forest products, leaving the challenge to make a distinction between uses for food and feed and for different conventional and new nonfood applications (bioenergy and bioproducts), at least at the level of regions and countries as import and export flows also need to be taken into account.

Land use modeling for predicting land use changes in response to the increasing demand for land has received increasing interest in recent years. This is related to increasing concerns about indirect land use changes (iLUC) caused by biomass use for energy leading to higher GHG emissions which may undo partly or completely the GHG mitigation effects of exchanging fossil fuels with biomass feedstock. One of the first published studies that raised general awareness of the size of additional GHG emissions that are caused by iLUC was from Searchinger et al. (2008).

Incorporation of the demand for fuels is well covered by several studies, certainly for reaching 2020–30 bioenergy targets of the EU, United States, Indonesia, Brazil, and sometimes other regions. Direct and indirect land use changes are assessed with the models in relation to satisfying demand for food, feed, and biomass for biofuels and wider bioenergy uses. At the same time these models generate data on biomass potentials as this must be part of the total modeling assessment. Examples of such modeling studies are presented in Annex 1.

Because of the many different models and studies performed there is no clear consensus on what the (direct and indirect) land use impacts are of nonfood biomass demand nor about the cost–supply functions of biomass for different nonfood uses. These differences in outcomes are first determined by the scenario specifications taken, particularly regarding sustainability constraints. Second, there are large differences in methodologies which relate to the model types, as discussed in the next section, and to the modeling approaches regarding important mechanisms such as yield developments, agricultural area expansion, and differences in input data, model calibration, and the way uncertainty is addressed. As to the latter aspect, this has improved a lot and has certainly helped to understand better the outcomes. In the most recent study assessing land use effects of EU biofuel targets, for example, which was performed with the GLOBIOM combination (Valin et al., 2016), extensive Monte Carlo sensitivity analysis was done with varying several influential factors determining the final outcomes on land use changes and related GHG emissions.

Most of these studies focus on assessing land use demands for agricultural resources and others combine the land use demand for agricultural and forest resources such as the case for the GLOBIOM model. However, studies incorporating the demand for biomass for electricity, heat, advanced biofuels, biochemical, and biomaterials are scarce, particularly because the demand for biomass for biochemical and biomaterials is not linked to policy targets or mandates (Schipfer et al., 2016). So the demand development is linked to technological development expectations and willingness to invest in the bio-based sectors (see also Chapter 5: Modeling Demand for Bio-Based Sectors).

The reason why the biomass cost-supply approaches in S2BIOM and Biomass Policies projects make use of the CAPRI and GLOBIOM models is related to the fact that these simulate land uses in agriculture and forestry (in case of GLOBIOM) very precisely. Both models take well account of the wide diversity in agricultural and forest products and the link to the diversified demand from a wide diversity of energy conversion options. CAPRI also takes account of high regional and farm diversity in Europe. GLOBIOM is very strong in integrating the agricultural, forest, and energy markets. Both CAPRI and GLOBIOM are partial equilibrium models. Therefore they do not cover the whole economy, but the sectors they do cover are represented in a very detailed way which makes their outcome from a land use perspective more reliable and useful for the assessment of biomass supply in the medium future.

2.3 MODEL TYPES AND MODELS USED IN S2BIOM AND BIOMASS POLICIES PROJECTS

In S2BIOM and Biomass Policies the main aim was to determine the biomass supply of the agricultural and forest sectors in Europe and to understand the cost-supply relation of main biomass imports into the EU until 2030. In this section it is explained why CAPRI and GLOBIOM models are very suitable to reach this aim.

To predict direct and indirect land demand effects, economic models are used to simulate the agricultural and timber (commodity) production in response to markets and trade relations between countries, and translate these land use changes (and if required) into GHG emissions. These economic models can be classified in computable general equilibrium (CGE) models and partial equilibrium (PE) models. The equilibrium refers to finding the equilibrium between supply and market demand for different sectors, depending on which are covered by the model. In these economic models, the principal pathways of market response to higher feedstock demand are higher prices and consequently reduced consumption, but also higher production through higher yields and ultimately an increase in the cultivated area and potentially a reduction in natural vegetation.

PE models focus on a subset of economic sectors and do not link explicitly to other sectors of the economy—e.g., the CAPRI and GLOBIOM models discussed in this chapter—and used to assess the biomass supply in the recent work performed within the S2Biom and Biomass Policies projects. The refined resolution in specific sectors important for biomass assessments; CAPRI covers the agricultural sectors in Europe and GLOBIOM covers the agricultural and forest sectors worldwide, allows incorporating a high level of detail for competition for inputs such as labor, water, energy, and fertilizer, and other detailed and complex relationships at high spatial and temporal resolution. This implies that the models cover smaller regional entities and make their simulations in small time steps allowing to take account is the large spatial diversity there and the interplay of many factors at the same time. However, the nature of PE models does not allow to consider links with other sectors which implies they can only be involved externally through input and output relationships with other models. A disadvantage is that economy-wide competition is beyond the scope of most PE models. This can introduce some bias when sectors have a big role in an economy.

CGE models include an entire economic system (national or global) by accounting for interactions between all productive sectors, labor, flow of goods and capital between sectors (or countries), and government policies. Most global and regional CGE frameworks are not structured to model land use alternatives and the associated emissions sources in detail.

In the following we provide a more extensive description of the functioning of the CAPRI and GLOBIOM models, input data, and links to other models.

2.3.1 CAPRI Model

CAPRI stands for Common Agricultural Policy Regionalised Impact. It was developed two decades ago with European Commission research funds and since then the model has been extended further geographically, in area coverage and in spatial detail in input and output data, methodologically, and in terms of EU policies covered. It is developed to evaluate ex ante impacts of the Common Agricultural and trade policies and of renewable energy policies on production, income, markets, trade, and the environment, from global to regional scale.

CAPRI is a global agricultural sector model with a focus on EU28, Norway, Turkey, and Western Balkans. The CAPRI model consists of two iteratively linked core modules (Fig. 2.1). The *supply module* covers around 280 (NUTS2) (see: http://ec.europa.eu/eurostat/documents/3859598/5916917/KS-RA-11-011-EN.PDF) regions and up to 10 farm types for each region. Each independent (regional) model maximizes regional agricultural income at given prices and subsidies, subject to constraints on land, policy variables, and feed and plant nutrient requirements. Most crop activities and permanent grasslands feature a high- and low-intensity variant to allow for endogenous yield adjustments. Yields and yield depending inputs are also updated based on price elasticities.

The *global market module* is a spatially comparative static multicommodity model for 47 primary and secondary agricultural products. It covers bilateral trade flows, including two way trade, based on the Armington's 1969 approach that considers commodities to differ by origin. Import prices take account of domestic prices of the exporter, transport cost, ad valorem and specific tariffs, export subsidies (subtracted), bilateral trade agreements, and Tariff Rate Quotas.

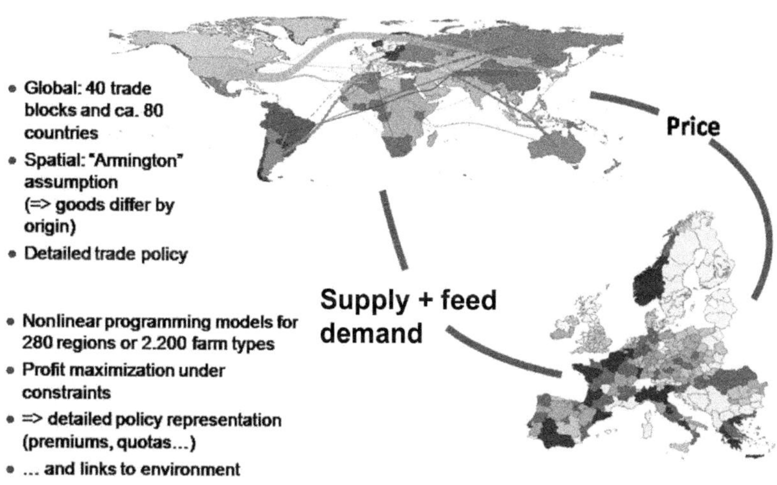

FIGURE 2.1 CAPRI model components and interactions.

In addition to the two core modules, there are supplementary modules and options that may be switched on or off depending on the issues under investigation. For example, a layer of *regional CGEs* has been developed in the CAPRI Rural Development project (see Törmä and Zawalinska, 2011; Britz, 2011).

CAPRI also includes a *land supply curve* which depends on the marginal return to land from the agricultural sector in each region and farm type. Substitution between permanent grassland and arable land is endogenous. In recent projects ("TRUSTEE", see: https://www.trustee-project.eu/), further steps are undertaken to extend the area coverage to nonagricultural UNFCCC land use categories and to cover agricultural GHG mitigation measures (Perez et al., 2016). The key benefit of CAPRI for Biomass Policies and S2BIOM has been however to provide a plausible agricultural baseline at NUTS2 level that is in line with projections by the PRIMES and GLOBIOM models.

2.3.1.1 Data Sources

The key database to the CAPRI modules is prepared in a module called "CoCo" (for Completeness and Consistency). This module integrates European, FAO, and OECD data sources but these raw data are consolidated to match balances and physical constraints. Missing data are estimated from associated time series and/or supplementary sources. National data are maintained and updated regularly. In addition to the most standard statistical input variables, there are also a wide number of variables included in relation to input coefficients per activity. Other database modules beyond CoCo estimate input coefficients per activity and region such that they completely exhaust the national sectoral costs shown in the Economic Accounts for Agriculture, while coming close to the Standard Gross Margins published by the European Commission.

2.3.1.2 Yield Changes

Yields and changes in activity levels per region and country in CAPRI are aligned in the medium-run perspective (up to 2025) with the AgLINK projections that also support the annual market projections by DG Agri. (https://ec.europa.eu/agriculture/sites/agriculture/files/markets-and-prices/medium-term-outlook/2016/2016-fullrep_en.pdf) The AgLINK system hosted at OECD takes information from questionnaires to all OECD Member States as a basis (see Britz and Witzke, 2011). The Member States fill in time series on future developments on several variables including yield developments of their main crops. The national input is then recovered in AgLINK by adapting the behavioral equations in the model while at the same time adapting these to joint worldwide future development expectations regarding import and export relations, worldwide price, and technological developments. CAPRI then takes AgLINK as an input, but further adapts future developments where needed to keep them within the ranges of analysis based on simple statistical trends for the different European States. These developments are then further incorporated into CAPRI but tuned where necessary with internal constraints set on yields for both vegetable and animal products.

2.3.1.3 Sectors Covered by CAPRI

The CAPRI model can now endogenously determine changes in supply and demand for fodder, food, and processing for biofuel feedstocks. As the market part of CAPRI includes behavioral functions for oilseed, and sugar and starch processing, the demand for biodiesel and bioethanol processing can be covered by either domestically or imported processed vegetable oils, and the domestic processing may be sourced from EU-produced feedstocks or imported ones. The following technology pathways are covered in CAPRI:

Total domestic ethanol production made from cereals (differentiated as wheat, barley, rye, oats, maize, and other), sugar, table wine, advanced (second-generation) ethanol (as far as the advanced fuel target share goes within the Renewable Energy Directive), and nonagricultural ethanol.

Biodiesel made from vegetable oils (differentiated as rape oil, sunflower oil, soya oil, and palm oil), advanced (second-generation) generation biodiesel, and nonagricultural biodiesel.

Advanced 2020 & 2030 biofuels produced from agricultural residues, like cereals straw or sugar beet leaves, or new energy crops (NERC), i.e., perennials.

In order to incorporate domestic use and supply of biofuels in the assessment, domestic ethanol and biodiesel production are defined using a profit-maximization approach as a function depending on processing margins. An exception is advanced (second-generation) and nonagricultural ethanol and biodiesel, for which production quantities are given exogenously and thus depend on baseline assumptions. The baseline assumptions leading to 2020 and 2030 projections are including expert knowledge, building on different sources that provide projections of domestic use and supply of biofuels for the individual EU28 countries (PRIMES) and non-European countries (AgLINK-COSIMO simulation results, see Blanco et al., 2010). For projections beyond 2020, GLOBIOM results are also considered with a weight that increases against AgLINK-COSIMO toward the long run. For the biofuel sector the key external information is PRIMES which in turn has assumed that EU biofuel targets are fully attained with a significant share based on second-generation technologies from lignocellulosic feedstocks. The latter biomass is coming from crop residues and NERC allocated in CAPRI.

2.3.2 GLOBIOM Model

The Global Biosphere Management (GLOBIOM) model (http://www.globiom.org/). (Havlík et al., 2014) is a global recursive dynamic PE model of the forest and agricultural sectors, where economic optimization is based on spatial equilibrium modeling (Takayama and Judge, 1971). The model projects developments on the forestry and agricultural markets, international trade, impacts on land use, and emissions and removals for the Agriculture, Forestry, and Other Land Use sector (AFOLU).

The model is based on a bottom-up approach where the supply side of the model is built up from the land cover, land use, and management systems at the bottom to production and markets at the top (Fig. 2.2). The agricultural and forest productivity is modeled at the level of grid cells of 5 × 5 to 30 × 30 minutes of arc, using biophysical models (G4M (www.iiasa.ac.at/G4M), EPIC (www.iiasa.ac.at/EPIC), RUMINANT (Havlik et al., 2014)), while the demand and international trade occur at

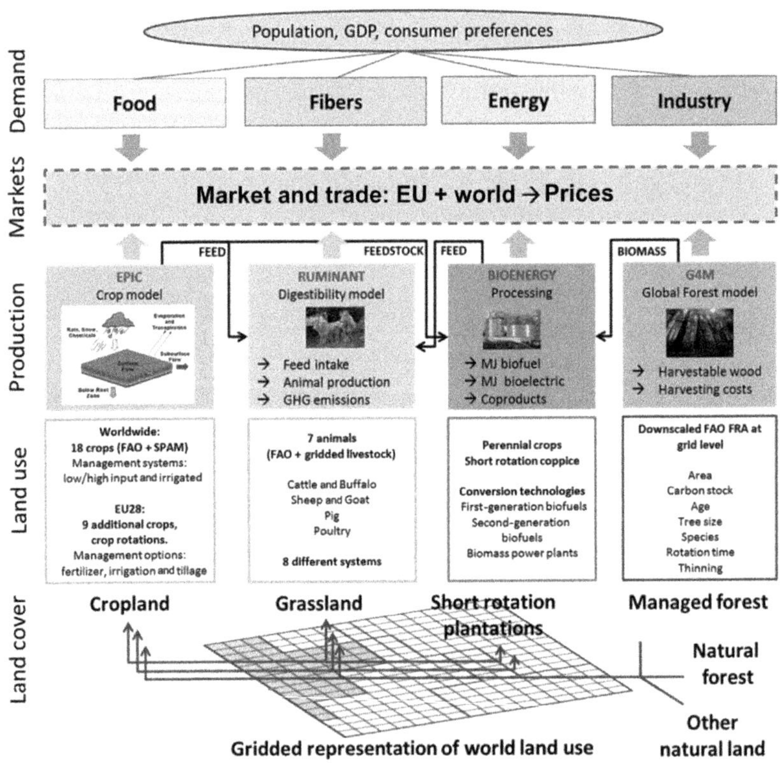

FIGURE 2.2 The GLOBIOM modeling framework.

regional level (30–53 regions covering the world, depending on the model version and research question). (The supply-side resolution is based on the concept of simulation units, which are aggregates of 5–30 arc-minute pixels belonging to the same country, altitude, slope, and soil class.) Besides primary products, the model has several final and by-products, for which the processing activities are defined.

Market equilibrium for forest and agriculture products is computed by allocating land use among production activities to maximize the sum of producer and consumer surplus, subject to resource, technological, and policy constraints. The level of production in a given area is determined by the forest or agriculture productivity in that area. Trade is modeled following the spatial equilibrium approach, which means that the trade flows are balanced out between different specific geographic regions. This allows tracking of bilateral trade flows between individual regions.

2.3.2.1 Representation of Land Use Change

The GLOBIOM model includes six land cover types (Fig. 2.3) and the economic activities are associated with cropland, grassland, plantations, and used and unused forests. (The term "used forests" refers to all forest areas where harvesting operations take place, while the term "unused forests" refers to undisturbed or primary forests. There are other three land cover types represented in the model to cover the total land area: other agricultural land, wetlands, and not relevant (bare areas, water bodies, snow and ice, and artificial surfaces). These three categories are currently kept constant at their

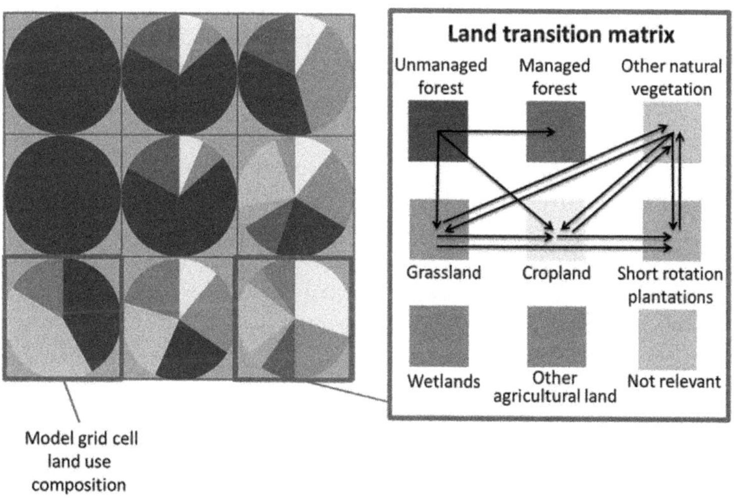

FIGURE 2.3 Land cover representation in GLOBIOM and matrix of endogenous land cover changes.

initial level.) Depending on the relative profitability of primary, by-, and final products' production activities, the model can switch from one land cover type to another. Initial land use is within the model set according to CORINE/PELCOM (http://land.copernicus.eu/pan-european/corine-land-cover) land cover estimates for EU and for the RoW the Global Land Cover 2000 (GLC 2000) is used. These sources of information provide an accurate baseline for the year 2000 according to which GLOBIOM can be calibrated. Land conversion over the simulation period is endogenously determined for each grid cell within the available land resources (see Fig. 2.3).

2.3.2.2 Representation of Agriculture Production and Livestock

GLOBIOM explicitly covers production of each of the 18 world major crops representing more than 70% of the total harvested area and 85% of the vegetal calorie supply as reported by FAOSTAT. Each crop can be produced under different management systems depending on their relative profitability: subsistence, low-input rainfed, high-input rainfed, and high-input irrigated, when water resources are available. Crop yields are generated at the grid cell level based on soil, slope, altitude, and climate information, using the EPIC model (www.iiasa.ac.at/EPIC). Within each management system, input structure is fixed following a Leontief production function. However, crop yields can change in reaction to external socioeconomic drivers and switch to another management system or reallocation of the production to a productive grid cell. Besides the endogenous mechanisms, an exogenous component representing long-term technological change is also considered.

The GLOBIOM model also incorporates a particularly detailed representation of the global livestock sector. With respect to animal species, distinction is made between dairy and other bovines, dairy and other sheep and goats, laying hens and broilers, and pigs. Livestock production activities are defined in several alternative production systems adapted from Seré and Steinfeld (1996) and Herrero et al. (2013). Feed rations in GLOBIOM are defined with a digestion model (RUMINANT, see Havlík et al., 2014) consisting of grass, stovers, feed crops aggregates, and other feedstuffs. The representation of the grass feed intake is an important component of the system representation as grazing land productivity is explicitly represented in the model. Therefore, the model can represent a full interdependency between grazing land and livestock.

2.3.2.3 Representation of Forests and Forest Industries

In terms of forestry, the model not only covers the development of the forest resources, but also projects the development of forest-based industries, and the use of woody biomass commodities for energy and material use. Total forest area in GLOBIOM is calibrated according to FAO Global Forest Resources Assessments (FRA) and divided into managed and unmanaged forest using a downscaling routine based on human activity impact on forest areas (Kindermann et al., 2008). (The term "unmanaged forests" refers to all forest areas that do currently not contribute to wood supply, based on economic decision rules in the model. However, they may still be a source for collection and production of nonwood goods (e.g., food, wild game, ornamental plants). Forests that are used in a certain period to meet the wood demand, the so-called managed forests, are modeled to be managed for woody biomass production. This implies a certain rotation time, thinning events, and final harvest.) G4M provides the available woody biomass resources in the form of mean annual increments for each forest area unit based on age class structural developments and adaptation of rotation periods and management to changes in demand. In addition to stem wood, available woody biomass resources also include branches and stumps; however, environmental and sustainability considerations constrain their availability and use for energy purposes.

Harvest costs for forests in GLOBIOM are based on G4M estimates through the use of spatially explicit constant unit costs that include planting, logging, and chipping in the case of logging residues. Transport costs are modeled through the use of regional level constant elasticity transport cost functions, which approximate the short-run availability of woody biomass in each region but are shifted over time in response to the changes in the harvested volumes and related investments in infrastructures.

The forest-based industries represented in the GLOBIOM model cover seven final products: chemical pulp, mechanical pulp, sawnwood, plywood, fiber board, other industrial round wood, and household fuelwood (Lauri et al., 2014). Forest industrial products are produced by Leontief production technologies and by-products of these technologies (bark, black liquor, sawdust, and saw chips) can be used for energy production or as raw material for pulp and fiberboard.

2.3.2.4 Representation of Short Rotation Coppices and Lignocellulosic Biomass

The land potentially available for the biomass production from Short Rotation Coppice (e.g., willow, poplar) and lignocellulosic perennial crops (e.g., miscanthus and reed canary grass) is defined in GLOBIOM based on land suitability and constraints that protect land use and land use change (e.g., Natura 2000, WDPA-protected areas). The use of agricultural land to produce short rotation crops (SRC) and lignocellulosic biomass is highly driven by the bioenergy demand, which is commonly set according to the PRIMES (Fragkos et al. (2017)) and POLES (Labat et al. (2015)) estimates. Cost structures are based on aggregated estimates of plantation and harvesting costs, while yields are currently based on NPP (Net Primary Productivity) maps and own estimats. For further details about cost and yield assumption, we refer to Havlík et al. (2011).

2.3.2.5 *Representation of Bioenergy Demand*

The GLOBIOM model also covers the demand of wood and agricultural biomass sources for energy purposes. Biomass use for energy production (heat, electricity, and biofuels) is commonly specified as a scenario specific input, based on the POLES, MESSAGE, or PRIMES energy sector models (Havlík et al., 2011; Reisinger et al., 2013; Fricko et al., 2016). The GLOBIOM model thereby projects which feedstocks will be used to fulfill the energy demand categories specified by the energy sector model. To fulfill the energy demand, GLOBIOM represents multiple conventional and advanced biofuels feedstocks, as well as production pathways for electricity and heat production (Havlík et al., 2011). This allows competition for biomass resources to be considered, also taking into account competition between the various sectors in terms of the demand for food, feed, timber, and energy.

2.3.2.6 *Categories of Biomass and Biomass Conversion Included in GLOBIOM*

GLOBIOM represents a number of conventional and advanced biofuels feedstocks. This allows ethanol, methanol, biodiesel, heat, electricity, and gas to be distinguished and traced according to their feedstocks. Furthermore, competition for biomass resources as considered is also taken into account between the various sectors in terms of the demand for food, feed, timber, and energy. The feedstocks that are covered include 27 different crops including four vegetable oil types (palm oil, rapeseed oil, soy oil, and sunflower oil); coproducts (three oilseed meal types, wheat and corn DDGS (Dried Distillers Grains with Solubles, DDGS)); perennials and short rotation plantations (miscanthus, switchgrass, Short Rotation Coppice); four types of stem wood, primary forestry residues from wood harvest; wood processing residues (see above); and recovered wood products and crop residues (e.g., straw).

Various energy conversion processes are modeled in GLOBIOM and implemented with specific technological costs, conversion efficiencies, and coproducts:

Wood (forestry): sawnwood, plywood, fiber board, pulp and paper production, combustion, fermentation, gasification

Lignocellulose (energy crop plantations): combustion, fermentation, gasification

Conventional ethanol: corn, sugarcane, sugar beet, and wheat ethanol processing

Conventional biodiesel: rapeseed oil, soybean oil, soya oil, and palm oil to FAME processing

Oilseed crushing activities: rapeseed, soybeans, and sunflower crushing activities.

2.3.2.7 Overview of Trade Representation in GLOBIOM

International trade of the considered feedstocks, processed, and final commodities from the forest, agriculture, and livestock sectors are computed endogenously within the GLOBIOM model between geographic regions. This implies that trade of commodities between EU28 and the RoW (including all non-EU28 countries) is endogenously estimated by the GLOBIOM modeling framework.

2.4 ASSESSMENT OF BIOMASS POTENTIALS FROM AGRICULTURE IN EUROPE USING THE CAPRI MODELING PLATFORM

For the assessment in this study land use and livestock production levels are used based on CAPRI projections for the 2013 reference scenario for DG ENER, TRANS, and CLIMA (Capros et al., 2013). Originally it covered the period from 2005 to 2050 in 5-year steps. For S2BIOM the CAPRI base year had been updated to 2008 and only the results for 2010, 2020, and 2030 have been used. The CAPRI baseline assumes compliance with EU policy regarding bioenergy targets based on the PRIMES energy model as reported in the *Trends to 2050 report* (Capros et al., 2013), including its effects on agriculture. The emphasis in the CAPRI reference has been on the food, feed, and biofuel crop and livestock at the national level and it provides detailed information on agricultural land use cropping and livestock patterns at regional (NUTS2) level. This implies that in a postmodel process (applied in Biomass Policies and S2BIOM), the CAPRI model output serves as the basis for estimating:

The potential for agricultural by-products in 2020 and 2030. The livestock patterns are also a good basis to calculate competing use levels for straw in livestock production.

The unused/released land potential in 2020 and 2030, compared with the base year of 2008, that may be used additionally for dedicated biomass cropping with perennial crops given expected increased demand for (lignocellulosic) biomass.

CAPRI model results still need further postmodel analysis to derive the maximum land availability for dedicated cropping from it. The reason is that the demand for lignocellulosic biomass from the wider bio-based economy, for biochemicals, bioheat, and bioelectricity, has so far not been fully internalized into CAPRI, yet such baseline results for lignocellulosic areas will not reflect a full use of the land potential.

In the following section we therefore explain how the output of the CAPRI baseline run was further used and processed in Biomass Policies and S2BIOM projects to come to biomass potential estimates for land availability estimates for dedicated perennial biomass crops. Detailed assumptions and indicators for the assessment of agricultural residues is discussed in Chapter 7, Assessing Potentials for Agricultural Residues.

2.4.1 Identification of Unused Lands in Europe Using CAPRI Baseline Results

From the CAPRI baseline run results, three types of land can be extracted that have a potential for dedicated biomass cropping which are not required for to satisfy the feed and food demand according to:

CAPRI:

1. Unused land, the land that was in agricultural use in 2008, but not in 2020 and 2030.
2. Land that is left fallow in 2020 and 2030 according to the CAPRI baseline.
3. Land is already dedicated to NECR according to the amount of advanced biofuels that are expected to be based on dedicated lignocellulosic crops as exogenously assessed with the PRIMES model (see earlier). These NECR lands are consistent with the demand for food, feed, and first-generation biofuel crop demands according to CAPRI.

The identification of the unused land requires the elaboration of a land use balance approach by comparing the land use situation in 2008 from CAPRI with 2020 and 2030. To identify, the following postmodel analysis steps were performed:

1. It was determined how much land in the arable land category was used in 2008 as compared to 2020 and 2030. If this land is larger in 2020 or 2030 there is no land released in this category and the land increase will need to come from losses in lands in the other land use categories, e.g., permanent grasslands and crops. If there is less land used we assume that there is a land release, which first needs to be (partly) absorbed for increase in permanent crops and permanent grassland categories. If there is still released land left after this, it may be counted as part of the land potential for dedicated biomass crops.
2. Next the land used in the permanent crop and the grassland categories for 2008 is compared with these lands in 2020 and 2030. Again it can be determined whether land is released or whether it has increased within these categories. If there are land releases they should first be used to absorb possible land use increases in the other categories, the remaining is potential for biomass crops.
3. Next the net land releases or increases are combined and if resulting in a net decline in agriculturally used land in 2020 and 2030 it can be regarded as "unused" by agriculture so far and thus potentially available

for (additional) dedicated biomass cropping. In addition fallow land and the land already used for NERC can be added to this resource.

4. The net land releases are distributed in two types of land: high- and medium- to low-quality lands. It is assumed that if there is a net land release in the arable or orchard category, this land is allocated to the high-quality group. Releases from olive groves, vineyards, and permanent grassland are allocated to the medium- to the low-quality group. Fallow land is assumed to be medium- to low-quality group. The land already allocated in CAPRI to NECR, given advanced biofuel target shares, is distributed 50% over good and 50% over medium to low quality (Table 2.1).

The land used for NERC for reaching advanced biofuel targets predicted by PRIMES and added exogenously to CAPRI (NECR) is only found in EU28 given that these countries have biofuel targets. These NECR lands make up already at least 60% of the land potential in EU28. In the non-EU countries the land category that dominates is fallow.

2.4.2 Land Availability Constrained by Sustainability Criteria

The land availability estimate derived from the postmodel assessment of CAPRI baseline is seen as a technical land potential. It is acknowledged that this net land potential cannot be regarded as entirely unused as there could still be other ecosystem service connected to these lands. This particularly applies to services like biodiversity and carbon sequestration. The Renewable Energy Directive (RED) criteria specified in Table 2.2 have been selected for constraining the technical land availability from the CAPRI baseline. Although currently the RED criteria are only applicable to biomass used to produce

TABLE 2.1 CAPRI Postmodel Analysis Results of Technical Land Potential Distributed Over High–Quality (HQ) and Low–Quality (LQ) Land (×1000 ha)

×1000 ha	HQ Land	LQ Land	Total
2020			
EU28	13688	4607	18295
Non-EU	11010	1760	12770
Total	24698	6367	31065
2030			
EU28	14459	5460	19919
Non-EU	10745	1760	12505
Total	25204	7220	32424

TABLE 2.2 RED Sustainability Criteria for Identifying Land Potential for Dedicated Biomass Crops

RED Criteria	Rules Implemented to Assess Land Availability and Selection of Suitable Woody and Perennial Crops	Technical Potential	Base Potential	User-Defined Potential
No loss of habitat of high biodiversity value	Exclusion of Natura 2000 and High Nature Value farmland areas		X	X
No use of areas of high carbon stock lands	Exclusion of: (1) wetlands and peatland areas/lands that have NOT been registered as agricultural since 1990 (ensures exclusion of continuous forest lands)/permanent grasslands (even if released from agriculture as assessed by CAPRI)	X	X	X
Avoidance of direct and indirect land use changes	Only use surplus (agricultural) lands	X	X	X
	Avoid use of Natura 2000 and HNV farmland (even if released from agriculture as assessed by CAPRI)		X	X
	Avoid conversion of permanent grasslands to arable		X	X
Support agro-biodiversity	No use of fallow land if fallow land share (in total arable land) declines to <10%			X
	Avoid monoculture choosing mix of at least three perennial crops per region (covering both woody and herbaceous crops)		X	X
Avoid negative impacts on soil quality and enhance soil quality impacts[a]	Maximum slope limits to perennial plantations[a]/use perennial plantations to protect soil susceptible to erosion[a] and for bio-remediation of polluted soils[a]		X	X
Avoid negative impacts on water resources[a]	Only use crops where minimal water requirement is delivered through annual precipitation (so irrigation is allowed but water depletion is avoided)[a]		X	X
	No use of irrigation in perennial crops[a] and preference for water use efficient crops in drought-prone regions[a]			X

[a]*These aspects of the RED Directive promote the use of certain crops and management practices. This aspect is further discussed in Chapter 9, Assessing the Potentials for Nonfood Crops.*

biofuels, in S2BIOM these are applied to all biomass used for nonfood applications. The application of certain RED criteria is dependent on the type of potential involved. In the technical potential the RED criteria are hardly relevant except for the inclusion of lands that have been registered as agricultural since 1990 and lands that are not used for productive activities. The user-defined potential is the most strictly restricted as it includes all restrictions applied in the base potential and additionally the use of fallow land below a 10% share of this land in the total arable land area and restricts irrigation in dedicated crops.

The way the RED criteria of Table 2.2 have been applied is explained in Table 2.3. The exclusion of areas of high biodiversity value and carbon-rich areas involves a spatial (geographic information system, GIS) analysis and the reduction was applied to all four land use categories in which the land releases have been classified, so also to the NERC and the fallow lands. After applying the RED criteria to the technical land availability, the

TABLE 2.3 No-Go Areas According to RED and Biophysical Constraints

No-Go Area Consideration per Scenario	RED and Biophysical Exclusion Criteria and Assumptions for Mapping	Source
High biodiversity lands are excluded from the land availability	A likeliness score for HNV farmland per region (NUTS2/3) was converted to an HNV farmland share for arable and permanent grasslands. This percent was excluded from the released land category derived from CAPRI	HNV farmland likeliness map: Paracchini et al. (2009)
Land released in the permanent grassland category cannot be used for dedicated cropping because of risk of soil carbon loss	CAPRI land use changes between 2008, 2012, 2020, 2030 can be tracked per land use type. Based on this balance it can be estimated how many permanent grassland areas go out of production between 2008 and 2012, 2020 and 2030, respectively These releases are not to be used for dedicated cropping	CAPRI land use change baseline results 2008, 2010, 2020, and 2030
Fallow land if the total fallow land share in arable land is below 10% (only applicable to user-defined potential)	Fallow land share in 2012, 2020, and 2030 can be calculated from CAPRI baseline per NUTS2 region by dividing the fallow land by the total arable land in different years. If the share is below 10% no fallow land can be used for dedicated cropping. If the fallow land share is above 10% the land up to 10% of fallow land share can be used for dedicated cropping	CAPRI land use change baseline results

TABLE 2.4 Land Availability ([a]million ha) in Technical, Base, and User-Defined Potential in 2020 and 2030

[a]1000 ha	Technical Potential	Base Potential	Strict Sustainability Potential
2020			
EU28	18.3	13.8	11.7
Non-EU[a]	12.8	6.4	5.7
Total	31.1	20.2	17.4
2030			
EU28	19.9	15.2	13.1
Non-EU[a]	12.5	6.3	5.6
Total	32.4	21.5	18.8

[a]Ukraine and Moldova are not covered by CAPRI. Land availability was determined by using the national agricultural statistics that register "unused farmlands" and fallow lands as a separate category. The 2012 data were kept constant toward 2020 and 2030.

sustainable land availability results, which is used in the base potential. From Table 2.4 it becomes clear that the available hectares are reduced by 40% in the base potential and 50% in the strict sustainability potential.

2.5 ASSESSMENT OF BIOMASS IMPORT POTENTIALS TO EU USING THE GLOBIOM MODELING FRAMEWORK

Within the S2BIOM project, the GLOBIOM model was used to assess biomass import potentials to the EU on the basis of macroeconomic assumptions (gross domestic product (GDP), population, diet patterns, etc.) and the utilization of land. On this base, the increasing trade of lignocellulosic biomass was evaluated and cost–supply curves created. For creating the curves the model was run with an added demand of biomass represented by a price (USD per ton of dry mass or USD per PJ) reflecting an arbitrary price that large-scale energy producers may be willing to pay for the feedstocks (in the case of bioenergy) or final commodity (in the case of biofuels). For 2020 and 2030, model results indicate the additional amount of biomass that can be supplied to a region by trade. By running the model for a range of biomass prices, the trade-related cost–supply curves were created.

The potential of import of lignocellulosic biomass was assessed and specified for biomass imported from countries outside Europe. One set of cost–supply curves was created to define how much biomass can be imported to Europe from the RoW for bioenergy purposes (mainly heat and electricity), and one curve defining the potential import of biofuels. Though the two cost–supply curves differ in terms of the end use of biomass that is imported, they are both defined in terms of the amount of biomass that could in the future be imported at a specific cost.

The assessment of trade potential mainly focused on a subsection of the resources from forestry and agriculture lands. This is because reliable data is difficult to get on a global scale for all resource categories and a number of the categories are not being traded due to their inherent characteristics, complicating their use for industrial purposes. The lignocellulosic biomass categories that were considered for the two cost–supply curves are:

Bioenergy:
Short rotation coppice and eucalyptus on agricultural land
Stemwood from thinning and final fellings
Stem and crown biomass from early thinnings
Wood pellets produced from industrial by-products, harvest residues, SRC or eucalyptus plantations
Industrial by-products (sawdust, wood chips, bark, etc.)
Biofuels:
Ethanol from woody biomass
Conventional ethanol from wheat, corn, sugarcane, and sugar beet
Conventional biodiesel from rapeseed oil, soybean oil, soya oil, and palm oil for FAME production
Oilseed from rapeseed, soybeans, and sunflower.

As GLOBIOM is an economic model, these trade-related cost–supply curves reflect the possibility in competition with the use of resources for other purposes, which contrasts with a "food/feed first" perspective that is also commonly utilized in these types of studies. Constraints were put in place to exclude from the analysis biomass from protected forests (as defined by the World Database on Protected Areas (WDPA Consortium, 2004).

2.5.1 Trade Assumptions Concerning Woody Biomass

As the representation of forest materials and the forest sector within the GLOBIOM model has been updated for the S2BIOM project, trade assumptions concerning the use of these new products have also been updated (Fig. 2.4).

Key underlying assumptions concerning trade for the assessment:

Wood pellets are within the model allowed to be produced from industrial by-products, SRC, and eucalyptus plantations. These conversion pathways are accounted for to allow for future developments and changes in pathways. Roundwood is modeled to not be eligible as a production feedstock for pellets due to the lack of globally consistent and reliable data of the current use of this source of wood for production of pellets.

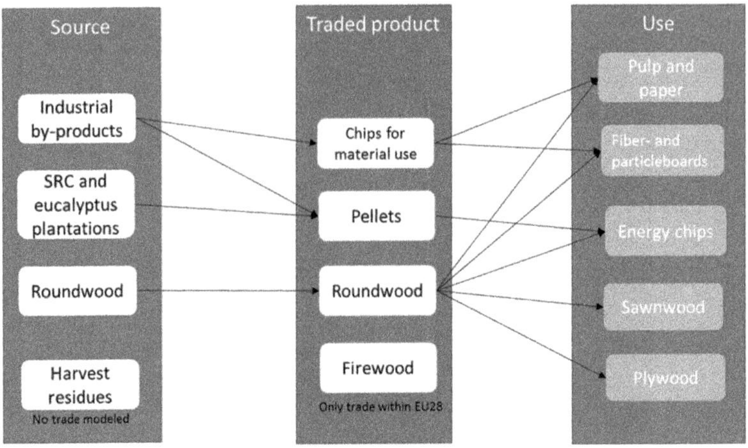

FIGURE 2.4 Trade of primary woody biomass between EU and the RoW as considered in GLOBIOM.

No trade of forest residues (Forest residues are typically leftover branches, stumps and stem tops from logging operations—thinning or final felling, chipped and mostly used for energy production. Forest residues are gathered from the logging site and forwarded to the roadside to be loaded on truck for long distance transport.) is modeled for this project as we currently do not consider this being traded due to a lack of reliable data to harmonize the trade flows.

Firewood (for household energy needs) is only allowed to be traded between EU countries as this is already common practice for small countries with small forest areas. Currently, there is some existing trade between non-EU countries, but its value is very small due to the bulkiness and low price (FAOstat).

Trade of wood chips for material and wood pellets for energy use is considered and accounted for within the GLOBIOM framework. It is noted that the majority of chips for the EU residential market is sourced locally, and imported chips are exclusively used in district heating or CHP plants (Pekkanen et al., 2014).

2.5.2 Main Underlying Assumptions for the Assessment

The following modeling features are furthermore reflected in the GLOBIOM integrated framework and utilized for this particular assessment:

Global population growth and GDP projections are exogenously assumed to develop over time. For the EU28, population and GDP are assumed to

develop fully in line with projections from the 2014 EU Impact Assessment (EC, 2014). Population growth and GDP projections for the RoW is assumed to develop according to the 2015 GECO POLES baseline scenario (Labat et al., 2015).
Only those sustainability criteria for biofuels, solid, and gaseous biomass that were assumed in the 2014 EU Impact Assessment (EC, 2014) are assumed to affect the development of the bioenergy sector and demand for bioenergy. These include (1) sustainable forest management practices that enhance forest productivity; (2) minimization of process chain emissions; and (3) efficient use of biomass to displace GHG-intensive fuels.
Demand for food and fiber is driven by human population growth and changing GDP. Demand for commodities is price elastic and therefore changes depending on consumers' willingness to pay. Demand is modeled through the use of constant elasticity functions which are parameterized by consumption quantities from EUROSTAT and FAOSTAT data on price and quantities. Historical data on production and prices of commodities (agricultural and forestry) was collected from individual country submissions, FAOSTAT, and EUROSTAT.
Bioenergy demand (heat, electricity, and biofuels) in the RoW evolves throughout the projection period in accordance with the baseline scenario of the 2015 GECO POLES report (Labat et al., 2015). Demands are implemented in the model as minimum constraints, meaning that a country can produce more but not less biomass for energy use than prescribed (e.g., not price elastic). It is thereby assumed that domestic bioenergy demand has to be fulfilled before exports to non-RoW countries can take place.
Household fuelwood (firewood) consumption for EU28 is assumed to develop over time according to PRIMES estimates of the development of the bioenergy production through small-scale solid conversion units. Data from the 2013 Joint Energy Wood Enquiry is used for the consumption levels for the year 2010, after which the use of firewood decreases if large-scale conversion units are installed and district heating networks are developed. Firewood consumption for the RoW strictly follows the POLES projections.

2.5.3 Estimated Cost–Supply Curves

The final cost–supply curves for bioenergy and biofuel purposes were estimated for 2010, 2020, and 2030 (Fig. 2.5), where the potential is shown in aggregate terms covering both import of wood chips and wood pellets. In terms of wood chips, the import to Europe is currently increasing following a global trend of increasing demand mainly for material purposes. Two

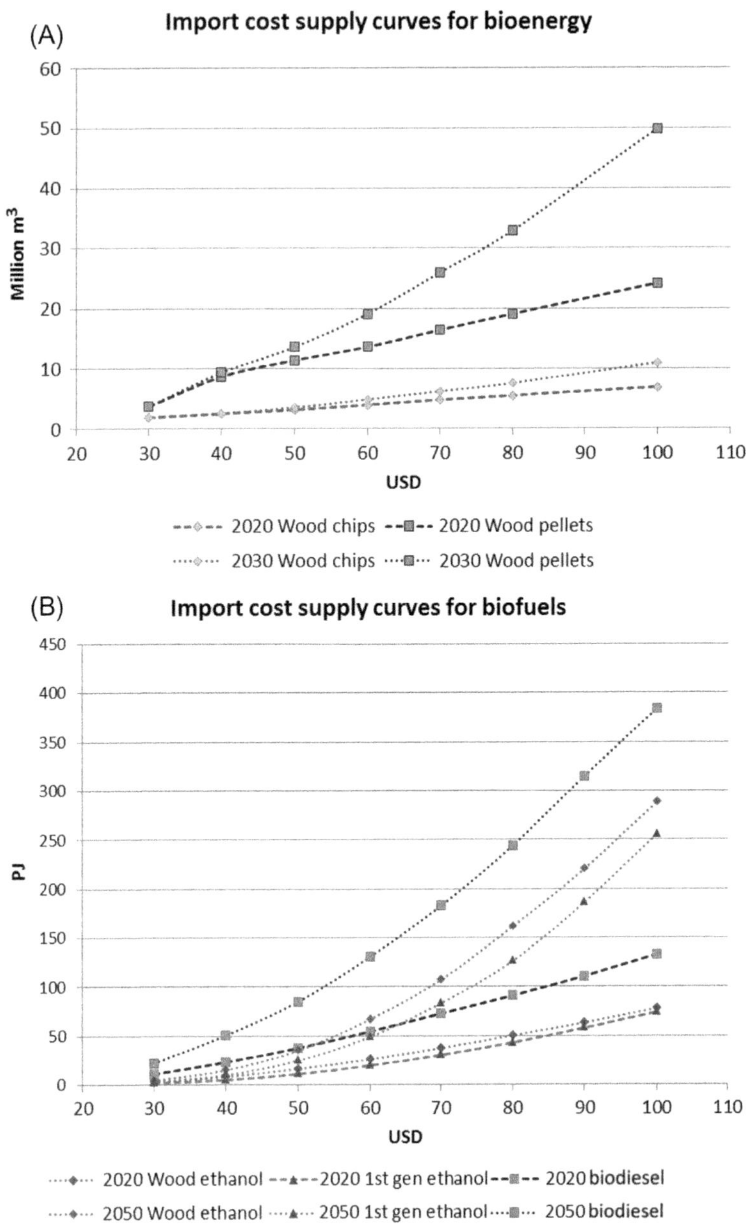

FIGURE 2.5 The estimated cost–supply curves in 2020 and 2030 for bioenergy (A) and biofuels (B). Note that the bioenergy potential is defined in terms of import of wood chips and wood pellets (million m^3), while the price is defined in terms of the cost that a consumer needs to pay for the feedstock (USD per m^3). Furthermore, the biofuel potential is defined in terms of import of first-generation ethanol from wheat, corn, and sugarcane; second-generation ethanol from woody biomass; and biodiesel from rape, sunflower, soya and palm oil (PJ). Price is defined in terms of the cost that a consumer needs to pay for the commodity (USD per PJ).

major import routes of woodchips to Europe can currently be identified. Hardwood chips are imported with sea vessels to Spain and Portugal mainly from Uruguay, Brazil, Canada, Congo, and Liberia. According to RISI (2012) the total Atlantic imports of woodchips to Spain and Portugal was about 2 million m^3. Another trade flow of wood chips originates mainly from Russia to Finland, amounting to 2.2 million m^3 (2012). It is important to note that these historically reported volumes do not separate the chips for pulp or wood-based panel production and energy production. It is estimated by a study from the IEA Bioenergy that less than 10% of the annually reported global wood chip trade volumes are energy-related. For wood chips for energy purposes, the main increase is estimated in terms of trade from Russia and northern parts of Africa.

The potential to increase import of wood pellets is more substantial than that of wood chips. Wood pellets are currently, by far, the most important solid wood fuel traded internationally. Demand in Europe draws currently the largest trade flows of wood pellets from North America and Russia. Import of wood pellets to Europe was about 3.2 million tonnes in 2012 and customs statistics show that the imports to the Europe in 2013 increased to 5.7 million tonnes. The US and Canada export about 2.7 and 2.0 million tonnes of pellets to Europe, respectively, and Russia about 0.7 million tonnes. The potential to increase in import of wood pellets to Europe is mostly expected for United States, Canada, and Russia. The imports from North America have been increasing during the last years and so far peak volumes 2.8 million tonnes from the United States and 1.9 million tonnes from Canada in 2013. During 2009–13 Russian exports to Europe have almost doubled from 0.4 million tonnes to 0.7 million tonnes.

2.6 CONCLUSIONS: MODEL LIMITATIONS AND RECOMMENDATIONS FOR FURTHER BIOMASS ASSESSMENT STUDIES IN THE CONTEXT OF MOST RECENT DEVELOPMENTS IN POLICY AND SOCIETY

Overall, it should be acknowledged that model outputs, even though coming from a well-documented and validated model, remain highly uncertain and they are always a simplification of the situation. Models cannot take all factors of influence into account, certainly if the future is far away.

For the postmodel evaluation of land availability with CAPRI, the mapping of no-go areas according to RED sustainability criteria course spatial data (low-resolution data) were used. The exact location of the unused land within every region for which CAPRI provided results was not known. The exact spatial match with no-go areas (according to RED) was not known and had to be estimated. An area share was assumed that matched with the regional area share, for typical soil and topographic characteristics and also overlapped with, for example, high nature value (HNV) farmland areas. So, if a share of the arable agricultural land was overlapping with HNV farmland, we assumed that the same area share applied to the released fallow and NERC lands. The results are likely to be less accurate for certain regions where unused land resources are unevenly spread within a region.

Part of the strength of the GLOBIOM model is that it is global and that countries and regions are explicitly linked through trade of commodities. Thus, the model can consider demand and supply developments on a global level and can assess the impact of national policies in terms of trade, leakage, and rebound effects. More importantly, it allows analysis of the EU biomass demand from a global perspective and projection of the competitiveness of EU feedstocks production and commodities on the international markets. The model thereby enables the assessment of the future share of EU domestic production where future import would originate, and the effect of such trade on importing as well as exporting countries. In addition, the GLOBIOM model is designed to integrate the forest and the agriculture sectors by representing both with a reasonable level of detail without compromising either, thus being a key model to assess the increasing interlinkages and feedbacks between these two sectors. While the forest-based industries were initially represented in a fairly simplified manner, key industrial sectors have been further disaggregated with great detail and the flow of material and by-products between individual industries has been improved (see Forsell et al., 2016; Rüter et al., 2016). This has improved the GLOBIOM model's capability to account for additional aspects such as the circular economy and capture the particularities of carbon implications of resource demand.

Overall, it may also be argued that agricultural and forestry feedstocks are too valuable to simply serve as sources of energy. From such a perspective biomass is a key feedstock for a whole bioeconomy that also involves energy use, but only at the endpoint of a whole cascade of proceeding uses. From this perspective the biomass potentials have not been addressed so far in both CAPRI, i.e., the CAPRI postmodel assessment and GLOBIOM.

The current specification of NERC modeling in CAPRI relies mostly on external information from PRIMES or elsewhere. This is due to the fact that the programming component of the CAPRI supply models includes, apart from observed accounting information, an unobserved and estimated cost function component to capture cost categories that are not explicitly handled in the model. These are labor and capital costs, risk components, crop rotation effects, and other factors supporting or counteracting specialization. All these factors are embedded in behavioral parameters (like price elasticities) that have been translated in terms of the CAPRI cost functions (see Jansson and Heckelei, 2010). However, to follow this approach, statistical observations are needed that are mostly absent for the emerging sector of NERC. In this situation a classical programming approach with a given cost function is an alternative that is also the key methodology underlying the whole PRIMES and GLOBIOM models. For CAPRI, the systematic data collection, control, and organization performed under S2BIOM (see also Chapter 9: Assessing the Potentials for Nonfood Crops) is a unique opportunity to specify the cost function for the difficult sector of NERC based on engineering information only. As labor and capital cost are covered in S2BIOM (see Chapter 9: Assessing the Potentials for Nonfood Crops) and the information is available even for NUTS3 regions of Europe, current efforts in ongoing projects are aiming exactly at such a specification of a cost function for dedicated bioenergy crops in CAPRI. At the same time the corresponding cost information for the collection of crop residues (straw and pruning residues) offers a similar option to also treat this type of biomass endogenously in CAPRI (see Chapter 7: Assessing Potentials for Agricultural Residues). This use of S2BIOM information in CAPRI benefits from the mappings between CAPRI and S2BIOM that have been established already in the early phase of the Biomass Policies and S2BIOM projects.

REFERENCES

Al-Riffai, P., Dimaranan, B., Laborde, D., 2010. Global Trade and Environmental Impact Study of the EU Biofuels Mandate. International Food Policy Research Institute, Washington, DC.

Blanco Fonseca, M., Burell, A, Gay, H., Henseler, M., Kavallari, A., M'Barek, R., et al., 2010. Impacts of the EU biofuel target on agricultural markets and land use: a comparatice modelling assessment. JRC Scientific Technical Reports. EUR 24449 EN – 2010.

Böttcher, H., et al., 2013. Future GHG emissions more efficiently controlled by land-use policies than by bioenergy sustainability criteria. Biofuels Bioproducts Biorefining Biofpr 7 (2), 115–125. Available from: <http://dx.doi.org/10.1002/bbb.1369>.

Britz W., 2011. RegCgeEU + , documentation including the Graphical User Interface, based on CAPPRI-RD Deliverable 3.2.4CAPRI. Website at: http://www.capri-model.org/dokuwiki/doku.php?.

Britz, W., Witzke, H.- P. (Eds.), 2011. *CAPRI Model Documentation 2008*: Version 2. Institute for Food and Resource Economics, University of Bonn, Bonn. Available from: <http://www.capri-model.org/docs/capri_documentation.pdf>.

Capros, P., De Vita, A.,Tasios, N., Papadopoulos, D., Siskos, P., Apostolaki, E., et al., 2013. EU energy, transport and GHG emissions trends to 2050—reference scenario 2013. <http://bookshop.europa.eu/en/eu-energy-transport-and-ghg-emissions-trends-to-2050-pbMJ0113672/;pgid = Iq1Ekni0.1lSR0OOK4MycO9B0000mrQD9vVA;sid = LkIsbsktO6YsO5whaIGyyasIjI FQPoROMDA = ?CatalogCategoryID = ydeep2IxOn0AAAEnSEMt7nca>.

European Commission, 2014. Impact assessment: accompanying the communication—a policy framework for climate and energy in the period from 2020 up to 2030.

Forsell, N., Korosuo, A., Havlík, P., Valin, H., Lauri, P., Gusti, M., et al., 2016. Study on impacts on resource efficiency of future EU demand for bioenergy (ReceBio). Final report. Project: ENV.F.1/ETU/2013/0033. Luxembourg, Publications Office of the European Union, 43 pp.

Fragkos, P., Tasios, N., Paroussos, L., Capros, P., Tsani, S., 2017. Energy system impacts and policy implications of the European Intended Nationally Determined Contribution and low-carbon pathway to 2050. Energy Policy 100, 216–226.

Fricko, O., Havlik, P., Rogelj, J., Klimont, Z., Gusti, M., Johnson, N., et al., 2017. The marker quantification of the shared socioeconomic pathway 2: a middle-of-the-road scenario for the 21st century. Global Environmental Change (http://pure.iiasa.ac.at/13326/1/The%20marker%20quantification%20of%20the%20Shared%20Socioeconomic%20Pathway%202.pdf).

Havlík, P., et al., 2011. Global land-use implications of first and second generation biofuel targets. Energ. Policy 39, 5690–5702.

Havlík, P., et al., 2014. Climate change mitigation through livestock system transitions. Proc. Natl. Acad. Sci. 111, 3709–3714.

Herrero, M., et al., 2013. Biomass use, production, feed efficiencies, and greenhouse gas emissions from global livestock systems. Proc. Natl. Acad. Sci. USA.<http://dx.doi.org/10.1073/pnas.1308149110>.

Hertel, T., Lee, H.-L., Rose, S., Sohngen, B., 2009. Modeling land-use related greenhouse gas sources and sinks and their mitigation potential. In: Hertel, T., Rose, S., Tol, R. (Eds.), Economic Analysis of Land Use in Global Climate Change Policy. Routledge Publishing.

Hertel, T.W., Golub, A.A., Jones, A.D., O'Hare, M., Plevin, R.J., Kammen, D.M., 2010. Global land use and greenhouse gas emissions impacts of U.S. maize ethanol: estimating market-mediated responses. BioScience 60, 223–231.

Jansson, T., Heckelei, T., 2009. A new estimator for trade costs and its small sample properties. Econ. Modell. 26 (2), 489–498.

Jansson, T., Heckelei, T., 2010. Estimating a primal model of regional crop supply in the European Union. J. Agric. Econ. vol 62 (nr 1), 137–152.

Kindermann, G.E., McCallum, I., Fritz, S., Obersteiner, M., 2008. A global forest growing stock, biomass and carbon map based on FAO statistics. Silva Fennica 42, 387.

Kraxner, F., et al., 2013. Global bioenergy scenarios—future forest development, land-use implications, and trade-offs. Biomass Bioenerg. 57, 86–96. Available from: <http://www.sciencedirect.com/science/article/pii/S0961953413000524>.

Labat, A., Kitous, A., Perry, M., Saveyn, B., Vandyck, T., Vrontisi, Z., 2015. GECO 2015. Global energy and climate outlook. Road to Paris. JRC scientific and policy reports, EUR 27239 EN.

Laborde, D., 2011. Assessing the land use change consequences of European biofuel policies. Final report ATLASS consortium. Contract no. TRADE/07/A2. Washington, DC, IFPRI. <http://trade.ec.europa.eu/doclib/docs/2011/october/tradoc_148289.pdf> (accessed 01.07.12).

Lauri, P., Havlík, P., Kindermann, G., Forsell, N., Böttcher, H., Obersteiner, M., 2014. Woody biomass energy potential in 2050. Energ. Policy 66, 19–31.

Melillo, J.M., et al., 2009. Indirect emissions from biofuels: how important? Science (New York, N.Y.) 326 (5958), 1397–1399. Available from: <http://globalchange.mit.edu/hold/restricted/MITJPSPGC_Reprint09-20.pdf>.

Moreira, M., Gurgel, A.C., Seabra, J.E.A., 2014. Life cycle greenhouse gas emissions of sugar cane renewable jet fuel. Environ. Sci. Technol. 48 (24), 14756–14763. Available from: <http://dx.doi.org/10.1021/es503217g>.

Mosnier, A., et al., 2013. Alternative U.S. biofuel mandates and global GHG emissions: the role of land use change, crop management and yield growth. Energ. Policy 57, 602–614. Available from: <http://linkinghub.elsevier.com/retrieve/pii/S0301421513001201>.

OECD/FAO, 2015. OECD-FAO Agricultural Outlook 2015, OECD Publishing, Paris. Available from: <http://dx.doi.org/10.1787/agr_outlook-2015-en>.

Oladosu, G., Kline, K., 2013. A dynamic simulation of the ILUC effects of biofuel use in the USA. Energ. Policy 61, 1127–1139. Available from: <http://dx.doi.org/10.1016/j.enpol.2013.06.124>.

Paracchini, M.L., Petersen, J.-E., Hoogeveen, Y., Bamps, C., Burfield, I., van Swaay, C., 2009. High nature value farmland in Europe. An estimate of the distribution patterns on the basis of land cover and biodiversity data. JRC Scientific and Technical Reports. European Communities, Luxembourg, <http://agrienv.jrc.ec.europa.eu/activities_HNV.htm>.

Pekkanen, M., Bowyer, C., Forsell, N., Hünecke, K., Korosuo, A., Nanni, S., et al., 2014. Study on Impacts on Resource Efficiency of Future EU Demand for Bioenergy (ReceBio), Task 1 Report. Project: ENV.F.1/ETU/2013/0033. Publications Office of the European Union, Luxembourg.

Pérez Domínguez, I., Fellmann, T., Weiss, F., Witzke, P., Barreiro-Hurlé, J., et al., 2016. An economic assessment of GHG mitigation policy options for EU agriculture (EcAMPA 2). JRC Science for Policy Report, EUR27973 EN. doi:10.2791/843461.

Plevin, R.J., et al., 2015. Carbon accounting and economic model uncertainty of emissions from biofuels-induced land use change. Environ. Sci. Technol. 49 (5), 2656–2664. Available from: <http://pubs.acs.org/doi/abs/10.1021/es505481d>.

Reisinger, A., Havlik, P., Riahi, K., van Vliet, O., Obersteiner, M., Herrero, M., 2013. Implications of alternative metrics for global mitigation costs and greenhouse gas emissions from agriculture. Climat. Change 117, 677–690.

REN21, 2012. Renewables 2012 global status report, renewables energy policy network for the 21st century. <http://www.ren21.net/REN21Activities/Publications/GlobalStatusReport/GSR2012/tabid/79218/Default.aspx>.

RISI International Pulpwood Trade Review 2012.

Rüter, S. Werner, F., Forsell, N., Prins, C., Vial, E. & Levet, AL., 2016. ClimWood2030 - climate benefits of material substitution by forest biomass and harvested wood products: perspective 2030. Final report. Thünen Report, No. 42. ISBN 978-3-86576-160-6, http://nbn-resolving.de/urn:nbn:de:gbv:253-201607-dn056927-3.

Schipfer, F., Lukas, K., Leclere, D., Sylvain, L., Forsell, N., Valin, H., 2016. Advanced biomaterials scenarios for the EU28 up to 2050 and their respective biomass demand. Biomass Bioenergy 96, 17–27.

Searchinger, T., Heimlich, R., Houghton, R.A., Dong, F., Elobeid, A., Fabiosa, J., et al., 2008. Use of U.S. croplands for biofuels increases greenhouse gases through emissions from land use change. Science 319 (5867), 1238–1240.

Seré, C., Steinfeld, H., 1996. World Livestock Production Systems: Current Status, Issues and Trends. Food Agriculture Organization, Rome.
Smeets, E., et al., 2017. Evaluating the land use change and food security effects of the use of residues and waste for bioenergy production. Biomass Bioenergy (submitted).
Takayama, T., Judge, G.G., 1971. Spatial and temporal price and allocation models.
Törmä, H., Zawalinska, K., 2011. Final documentation of the CGERegEU + model, CAPPRI-RD Deliverable 3.2.3.
Valin, H., Peters, D., van den, Berg, M., Frank, S., Havlik, P., Forsell, N., et al., 2016. The land use change impact of biofuels consumed in the EU, quantification of area and greenhouse gas impacts. Report prepared by Ecofys/IIASA/E4tech for DG ENER under contract BIENL13120.
Verburg, R., Stehfest, E., Woltjer, G., Eickhout, B., 2009. The effect of agricultural trade liberalisation on land-use related green house gas emissions. Global Environ. Change 19 (4), 434–446.
Verstegen, J.A. et al., 2016. What can and can't we say about indirect land use change in Brazil using an integrated economic—land use change model? GCB Bioenergy.
WDPA Consortium, 2004. World Database on Protected Areas. Copyright World Conservation Union (IUCN) and UNEP—World Conservation Monitoring Centre (UNEP-WCMC).

FURTHER READING

Buongiorno, J., Zhu, S., Zhang, D., Turner, J., Tomberlin, D., 2003. The global forest products model: structure, estimation, and applications. Academic Press, San Diego, CA.
Carpentieri, A.E., Larson, E.D., Woods, J., 1993. Future biomass-based electricity supply in Northeast Brazil. Biomass Bioenergy 4, 149–173.
Dees, M., Elbersen, B., Fitzgerald, J., Vis, M., Anttila, P., Forsell, N., et al., 2016. A spatial data base on sustainable biomass cost–supply of lignocellulosic biomass in Europe—methods & data sources. Project report. S2BIOM—a project funded under the European Union 7th Frame Programme. Grant Agreement no. 608622. University of Freiburg, 170 pp.
Fontagné, L., Gaulier, G., Zignago, S., 2008. Specialization across varieties and north–south competition. Econ. Policy 2008, 51–91.
FRA, FAO. Global Forest Resources Assessment, 2010. Main report.
Herzogbaum GmbH, 2008. Forstpflanzen-Preisliste 2008. HERZOG.BAUM Samen & Pflanzen GmbH. Koaserbauerstr. 10, A-4810 Gmunden, Austria. Available from: <www.energiehoelzer.atS>.
ILO, 2007. Occupational wages and hours of work and retail food prices, statistics from the ILO October inquiry. International Labor Organisation, ISBN: 9220201747. Available from: <http://laborsta.ilo.org/S>.
Jurve, M., 1997. Labor-intensive harvesting of tree plantations in the southern Philippines. Forest harvesting case-study 9. RAP publication: 1997/41, Food and Agriculture Organization of the United Nations.
Robinson, T.P., et al., 2011. Global Livestock Production Systems. Food Agriculture Organization; International Livestock Research Institute, Rome.
Seale, J.L., Regmi, A., Bernstein, J., 2003. International Evidence on Food Consumption Patterns. Economic Research Service, US Department of Agriculture, Washington, DC.

ANNEX 1 OVERVIEW OF MAIN MODEL CHARACTERISTICS FOR SELECTION OF MODELS USED TO PREDICT LAND USE CHANGES

Overview of Modelling Approaches Assessing Direct and Indirect Land Use Changes to Satisfy Demand for Food and Biomass for Nonfood Applications

Model	Type of Model PE/CGM	Main Sectors Covered	Geographic Coverage	Main Model Output/Focus	References for Model Applications in Studies	Model Holder
MIRAGE	Computable general equilibrium model, recursive dynamic	Multisectoral model up to 57 sectors	World divided in 113 regions	Effects of any policy, particularly agricultural, trade and biofuels policies on trade, capital, agricultural, and labour markets	Laborde et al. (2011) and Al Riffai et al. (2010)	IFPRI, INRA, developed initially at CEPII
GTAP	Global Trade Analysis Program (GTAP): computable general equilibrium model	Multisectoral model	Global	Global economic analysis of policy effects in relation to energy, land use and land cover change, poverty, labour migration, GHG emissions (EPPA-model link)	Hertel et al. (2009, 2010); Melillo et al. (2009); Oladosu and Kline (2013); Moreira et al. (2014); and Plevin et al. (2015)	GTAP consortium coordinated by Purdue University, OECD, USDA, IFPRI, WER, EPA, EC, Thunen Institute, etc.
GLOBIOM	Global Biosphere Management Model. Global recursive dynamic partial equilibrium (PE) model	Integrating the agricultural, bioenergy, and forestry sectors	Global; 50 world regions of which 27 are the individual EU Member States	To provide policy analysis on global issues concerning land use competition between the major land-based production sectors (LUC and iLUC) and GHG emissions (link EPIC model)	Havlík et al. (2014); Havlík et al. (2011); Forsell et al. (2016); Mosnier et al. (2013); Böttcher et al. (2013); Kraxner et al. (2013); and Valin et al. (2016)	IIASA developed it and is main user
MAGNET	Modular Applied GeNeral Equilibrium Tool (MAGNET) is a global general equilibrium model	Integrating all (57) sectors of the economy (agriculture, manufacturing, and services)	Global; divided in 112 regions	MAGNET is Modular set-up around a core of GTAP models. Main applications for assessing policy changes in agricultural, trade, and bioenergy on land use, trade, markets, labor, food security	Verstegen et al. (2016); Smeets et al. (2017); and Verburg et al. (2009)	Developed by Agricultural Economic Institute The Hague, now MAGNET consortium including JRC-IPTS, Thunen Institute

(*Continued*)

(Continued)

Model	Type of Model PE/CGM	Main Sectors Covered	Geographic Coverage	Main Model Output/Focus	References for Model Applications in Studies	Model Holder
CAPRI	Common Agricultural Policy Regionalized Impact. Partial equilibrium model, nonlinear	Agricultural sector highly detailed in terms of regional and farm diversity	CAPRI is a global agricultural sector model with a focus on EU28, Norway, Turkey, and Western Balkans	It is developed to evaluate ex-ante impacts of the common agricultural and trade policies and of renewable energy policies on production, income, markets, trade, and the environment from global to regional scale	Jansson and Heckelei (2009, 2010), and Pérez Domínguez et al. (2016)	It was developed by Uni. Bonn with EC research funds. Many users all over Europe
AGLINK-COSIMO	Recursive-dynamic, partial equilibrium model	AGLINK; 10 OECD countries and regions (Australia, Canada, European Union (EU), Switzerland, Norway, Japan, Korea, Mexico, New Zealand and the US) and 4 non-OECD countries (Argentina, Brazil, China, and Russia). The EU module consists of the 28 MS. Cosimo: 42 regions; 3 OECD members (Chile, Israel, and Turkey), 27 single countries, and 12 regional aggregates		Model analyses supply and demand of world agriculture. Simulates developments of annual market balances and prices for the main agricultural commodities produced, consumed, and traded worldwide	OECD/FAO (2015)	OECD, FAO

Chapter 3

Lignocellulosic Biomass Quality: Matching Characteristics With Biomass Conversion Requirements

Wolter Elbersen, Tijs M. Lammens, Eija A. Alakangas, Bert Annevelink, Paulien Harmsen and Berien Elbersen

3.1 INTRODUCTION TO LIGNOCELLULOSIC BIOMASS CONVERSION OPTIONS AND BIOMASS TYPES

3.1.1 Introduction to Biomass Quality and Conversion Systems

A wide range of conversion options for lignocellulosic biomass exist, or are under development that can convert biomass into bio-based products, chemicals, fuels, power, and heat.

We can divide these conversion technologies into three main categories:

1. Thermal conversion technologies
2. Biochemical conversion technologies
3. Anaerobic digestion.

Each of these conversion technology options has specific requirements for the biomass used as a feedstock. A set of the most relevant criteria can be defined which the biomass needs to meet, in order to be processed by a specific technology. Thermal conversion technologies have criteria for corrosion, ash melting behavior (agglomeration, fouling), ash content, and NO_x emissions. Chemical and biochemical processes have criteria on the lignin, (hemi)cellulose, and ash content. Anaerobic digestion has criteria for digestibility and biogas yield. In Section 3.2 these requirements are discussed in more detail.

Lignocellulosic biomass consists mostly of lignin, cellulose, and hemicellulose, and to a more or lesser extent other plant components such as sugars, starch, acids, fats, and oils, and last but not least water (moisture) and ash. Quality is essentially a set of biomass characteristics that determine the value of that biomass type for a certain conversion system. Some of these

Modeling and Optimization of Biomass Supply Chains.
DOI: http://dx.doi.org/10.1016/B978-0-12-812303-4.00003-3

characteristics are "fundamental" biomass characteristics, because they cannot easily be changed and are determined by the plant itself. Others are not so fundamental and can be adjusted easily. Particle size, moisture content, and bulk density are examples. Adjustments can be made with basic pretreatment technologies such as drying, compacting (i.e., pelletizing), or size reduction (i.e., chipping or crushing). These characteristics are generally dealt within postharvest logistical chains and discussed in Chapter 4, Modeling Biomass Logistics.

Other characteristics may be a result of the methods used in harvesting and handling. For example, contamination with soil during harvest, which can increase ash content dramatically, making the biomass useless for most conversion systems or requiring costly cleaning operations. Adapting harvest and handling can solve these problems. More fundamental biomass characteristics can be much harder to change. Examples are lignin content, ash content, or chlorine and potassium contents. If a more fundamental biomass characteristic does not match to the conversion system requirements, it may be better to find other biomass for that conversion system or process that biomass in another conversion system that can better handle this type of biomass.

In Section 3.3 the biomass or plant characteristics that determine quality conversion systems are discussed and explained, including options to manipulate them before harvest. Section 3.4 describes a methodology to match biomass characteristics to conversion demands, to help assess if the biomass matches conversion system requirements.

3.2 BIOMASS CONVERSION OPTIONS AND THEIR REQUIREMENTS

To match biomass to technologies we need to identify the fundamental biomass characteristics that determine if certain biomass can be used in a conversion technology and to what extent it matches. Fig. 3.1 illustrates the

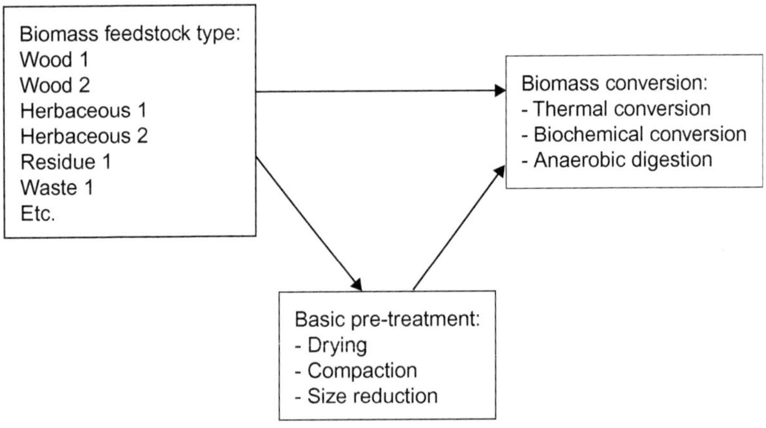

FIGURE 3.1 Illustration of the process of matching biomass types to biomass conversion systems with or without basic pre-treatments.

TABLE 3.1 Fundamental Biomass Properties Relevant for Thermal Conversion, Biological Conversion and Anaerobic Digestion

Property	Unit	Classification				Relevant Conversion Technology?
		1	2	3	4	
Thermal Conversion						
Chlorine content	wt% d.m.	<0.02	0.02–0.1	0.1–0.4	>0.4	Thermal conversion
Ash DT	°C	>1200	1000–1200	800–1000	<800	Thermal conversion
Ash content	wt% d.m.	<1	1–3	3–10	>10	Thermal and biochemical conversion
Nitrogen content	wt% d.m.	<0.3	0.3–1	1–2.5	>2.5	Thermal conversion
Biological Conversion						
Carbohydrates	wt% d.m.	>65	50–65	30–50	<30	Biochemical conversion
Lignin content	wt% d.m.	<10	10–25	25–35	>35	Biochemical and anaerobic digestion
Anaerobic Digestion						
Biogas yield	m^3/ton a.r.	>300	150–300	50–150	<50	Anaerobic digestion
Does the digestate have an application?		Yes			No	Anaerobic digestion

wt% d.m, weight percentage dry matter (dry basis); a.r., as received (wet basis).

matching concept. The fundamental biomass characteristics that determine the value or risks for a type of conversion system were identified, and classified into four ranges or classes, with class 1 as the "most desirable" class (e.g., low ash content, high biogas yield, etc.), and class 4 the "least desirable". Table 3.1 provides the ranges per class for each property, which are explained as follows.

3.2.1 Thermal Conversion

For a thermal conversion system, the main feedstock-related challenges relate to ash, nitrogen, chloride, and ash melting behavior, which are further discussed in Section 3.3.5.

3.2.2 Biochemical Conversion

A (bio-)chemical conversion system for lignocellulosic biomass will generally consist of a pretreatment step to release the C6 and C5 sugars followed by conversion of the polysaccharides into products like fuels or chemicals. Second-generation cellulosic ethanol production is the best known example.

Fundamental biomass characteristics that influence the potential success of biochemical conversion of lignocellulose biomass into fuels and chemicals are:

- *Lignin content*: Lignin can hardly be degraded by enzymes and microorganisms. Secondly, it acts as a shield that prohibits the bioconversion of cellulose and hemicellulose. The higher the lignin content, the more difficult it is to use lignocellulose as feedstock in biochemical conversion processes. Lignin is also associated with formation of antiquality compounds or inhibitors, such as polyphenols. These may hinder downstream fermentation processes.
- *High carbohydrate content*: The main components of interest in biochemical conversion of (lignocellulosic) biomass are cellulose and hemicellulose, which are carbohydrate polymers consisting of sugar polymers. High cellulose and hemicellulose content will make biomass more suitable for biochemical conversion.
- *Low ash content*: As with the other two conversion methods, ash or inorganic material cannot be converted within biochemical processes and generally adds to the costs of conversion.
- *Protein*: This may also be relevant from another point of view. Protein could be extracted and valorized so a higher content could be desirable if the protein can be removed.

3.2.3 Anaerobic Digestion

In aerobic digestion a large part of the cost is determined by the size of the reactor. So it is important that the yield per reactor volume per year is high enough. Therefore, the methane yield per ton (or m^3) of substrate is very relevant. The cost of the disposal of the residue (digestate) is also relevant, the potential applicability of the digestate (e.g., as fertilizer) should be taken into account. Local or national regulations determine if the digestate can be applied at a low cost or not. The biogas yield for biomass depends on the composition (ash, cellulose, hemicellulose, lignin, fat, protein, etc.). Ash and lignin will not contribute to biogas formation and cellulose will only contribute if it can be broken down at a sufficiently high rate. For straw-type material the biogas yield will be 0.15–0.5 m^3 biogas per kg volatile solids (Seadi et al., 2008).

3.3 BIOMASS QUALITY CHARACTERISTICS

3.3.1 Introduction

The two main types of lignocellulosic biomass that can be distinguished are herbaceous biomass and woody biomass. Within these two types, many different categories can be identified based on the standard classification of EN ISO 17225-1:2014 classification system (see Annex 1). This standard is used to define different categories relevant for thermal conversion and other biomass conversion systems.

3.3.2 Herbaceous Biomass

Herbaceous biomass is from plants that have a nonwoody stem and which die back at the end of the growing season. This biomass includes most agricultural crops and grasses, including bamboo and wheat straw (see EN ISO 17225-1:2014; Alakangas et al., 2016). Also relatively young and essentially nonwoody parts of trees exhibit similar characteristics. In general, herbaceous biomass will have higher nutrient contents and lower lignin contents than wood. Herbaceous biomass is variable in composition depending on the time of year and on the type of tissue. Composition can also be strongly influenced by availability of minerals or nutrients in the soil. The classification system for lignocellulosic biomass is given in Tables 3.A.1–3.A.8 in Annex 1. The categories based on this classification are discussed in the following sections.

3.3.3 Energy Grasses, Annual and Perennial Crops (Classification Number 211)

Perennial biomass grasses like switchgrass (*Panicum virgatum*), *Miscanthus* (*gigantheus*), reed canary grass (*Phalaris arundinacea*), giant reed (*Arundo donax*), and reed (*Phragmites australis*) differ in their characteristics depending on harvest time. Data on characteristics are generally taken in fall or after a killing frost in winter or early spring. In fall K, Cl, Na, and ash content will be significantly higher than in winter. The differences are due to loss of leaves (having a higher ash content than the stalk) during winter and leaching of components that are easily dissolved in water such as Cl, K, and Na. This explains the large differences in K and Cl contents reported for these types of biomass. Therefore, it would be logical to present fall and winter/spring-harvested data separately. However, this is difficult as most data sources of biomass composition do not clearly define the time of harvest. Fall-harvested grasses will have higher K, Cl, and ash contents and lower ash melting points than winter/spring-harvested grasses which should have lower K, Cl, and ash contents and a higher ash melting point. Still, herbaceous biomass will almost always have lower ash deformation temperature

TABLE 3.A.1 Subcategories of First Level Category 1 "Forestry"

Second Level Subcategories		Third Level Subcategories		Final Level Subcategories	
ID	Name	ID	Name	ID	Name
11	Production from forests	111	Stemwood from final fellings and thinnings	1111	Stemwood from final fellings originating from nonconifer trees
				1112	Stemwood from final fellings originating from conifer trees
				1113	Stemwood from thinnings originating from nonconifer trees
				1114	Stemwood from thinnings originating from conifer trees
12	Primary residues from forests	121	Logging residues from final fellings and thinnings	1211	Logging residues from final fellings from nonconifer trees
				1212	Logging residues from final fellings from conifer trees
				1213	Logging residues from thinnings from nonconifer trees
				1214	Logging residues from thinnings from conifer trees
		122	Stumps from final fellings and thinnings	1221	Stumps from final fellings originating from nonconifer trees
				1222	Stumps from final fellings originating from conifer trees
				1223	Stumps from thinnings originating from nonconifer trees
				1224	Stumps from thinnings originating from conifer trees

TABLE 3.A.2 Subcategories of Second Level Category "21 Primary Production of Lignocellulosic Biomass Crops"

Third Level Subcategories		Final Level Subcategories	
ID	Name	ID	Name
211	Energy grasses, annual, and perennial crops	2111	Sweet and biomass sorghum (annual grasses)
		2112	*Miscanthus* (perennial grass)
		2113	Switchgrass (perennial grass)
		2114	Giant reed (perennial grass)
		2115	Cardoon (perennial crop)
		2116	Reed canary grass (perennial grass)
212	Short rotation coppice	2121	SRC willow
		2122	SRC poplar
		2123	Other SRC

TABLE 3.A.3 Subcategories of Second Level Category "22 Agricultural Residues"

Third Level Subcategories		Final Level Subcategories	
ID	Name	ID	Name
221	Straw/stubbles	2211	Rice straw
		2212	Cereals straw
		2213	Oil seed rape straw
		2214	Maize stover
		2215	Sugarbeet leaves
		2216	Sunflower straw
222	Woody prunings and orchard residues	2221	Residues from vineyards
		2222	Residues from fruit tree plantations (apples, pears, and soft fruit)
		2223	Residues from olives tree plantations
		2224	Residues from citrus tree plantations
		2225	Residues from nuts plantations

TABLE 3.A.4 Subcategories of Second Level Category "23 Grassland"

Third Level Subcategories		Final Level Subcategories	
ID	Name	ID	Name
231	Grassland	2311	Unused grassland cuttings (abandoned grassland, managed grasslands not used for feed)

TABLE 3.A.5 Subcategories of "3 Other Land Use"

Third Level Subcategories		Final Level Subcategories	
ID	Name	ID	Name
311	Biomass from other areas under landscape maintenance	3111	Grassy biomass from landscape maintenance (recreational and nature protection areas, dykes)
		3112	Woody biomass from landscape maintenance (landscape elements)
312	Biomass from roadside verges	3121	Grassy biomass from roadside verges
		3122	Woody biomass from roadside verges

TABLE 3.A.6 Subcategories of Second Level Category 41 "Secondary Residues From Wood Industries"

Third Level Subcategories		Final Level Subcategories	
ID	Name	ID	Name
411	Sawmill residues	4111	Sawdust from sawmills from conifers
		4112	Sawdust from sawmills from nonconifers
		4113	Sawmill residues: excluding sawdust, conifers
		4114	Sawmill residues: excluding sawdust, nonconifers
412	Other wood processing industry residues	4121	Residues from industries producing semifinished wood-based panels
		4122	Residues from further wood processing
413	Secondary residues from pulp and paper industry	4131	Bark residues from pulp and paper industry
		4132	Black liquor

TABLE 3.A.7 Subcategories of "42 Secondary Residues of Industry Utilizing Agricultural Products"

Third Level Subcategories		Final Level Subcategories	
ID	Name	ID	Name
421	By-products and residues from food and fruit processing industry	4211	Olive stones
		4212	Other by-products and residues from food and fruit processing industry
422	Other industry by-products utilizing agricultural products	4221	Cotton acorn
		4222	Other industry by-products utilizing agricultural products

TABLE 3.A.8 Subcategories of "Waste"

Second to Third Level Subcategories		Final Level Subcategories	
ID	Name	ID	Name
51/511	Biodegradable municipal waste	5111	Biowaste as part of integrally collected municipal waste: Biodegradable waste of not separately collected municipal waste (excluding textile and paper)
		5112	Separately collected biowaste: Biodegradable waste of separately collected municipal waste (excluding textile and paper)
52/521	Postconsumer wood	5211	Hazardous postconsumer wood
		5212	Nonhazardous postconsumer wood

(DT) than woody biomass (Alakangas et al., 2016) and therefore higher risk of slagging and fouling. A large difference in biogas yield may also be expected as green material will have a higher yield than dry material harvested in winter.

3.3.4 Woody Biomass

Woody biomass is a very large category of biomass in total available volume and in subcategories. Wood is essentially dead plant material containing thick cell walls with a high lignin content. Wood has very good characteristics for thermal conversion mainly due to the very low ash, N, K, Na, and Cl contents resulting in a high ash melting point and low risk of corrosion compared to herbaceous biomass. The categories are further described in the following sections.

3.3.4.1 *Primary Forestry Residues (Classification Number 12)*

Especially for broadleaf trees, but also for conifers, the moisture content decreases with the dimension of the stem.

3.3.4.2 *Sawdust (Conifers and Nonconifers) (Classification Numbers 4111 and 4112)*

For lignin–cellulose–hemicellulose content, the same was assumed as stemwood from final fellings. Sawdust usually does not include bark, so ash content is lower than stemwood or thinning wood.

3.3.4.3 *Residues From Industries (Classification Number 412)*

These residues might also include some chemicals (e.g., glue) so that it is not possible to assume a composition similar to virgin wood, e.g., residues from plywood industry, furniture industry, and particle board industry. If these residues do not include any chemicals, e.g., sawdust, cutter chips, and shavings, then the composition is similar to that of stemwood.

3.3.4.4 *Black Liquor (Classification Number 4132)*

Black liquor is a by-product of paper pulp production mostly consisting of the lignin part removed during chemical pulping. It usually has a dry matter content of approximately 80 wt% (Alakangas et al., 2016). The dry matter content increases when the recovery boiler size increases.

3.3.5 Explanation of Biomass Characteristics

This section discusses the main biomass characteristics relevant for conversion explaining typical sources of variation in composition (Fig. 3.2).

3.3.5.1 *Moisture Content*

The risk of spoilage, biomass loss, and self-ignition increases with a higher moisture content. For herbaceous biomass efficient storage and transport is not possible, if moisture is above 15% (on a wet basis). For wood this value is 23% for small-scale use. In Nordic countries, when outdoor storage of logging residues and thinning wood is the prevailing method, the storage moisture is higher, usually about 35–45 wt% on wet basis. During storage, moisture content decreases.

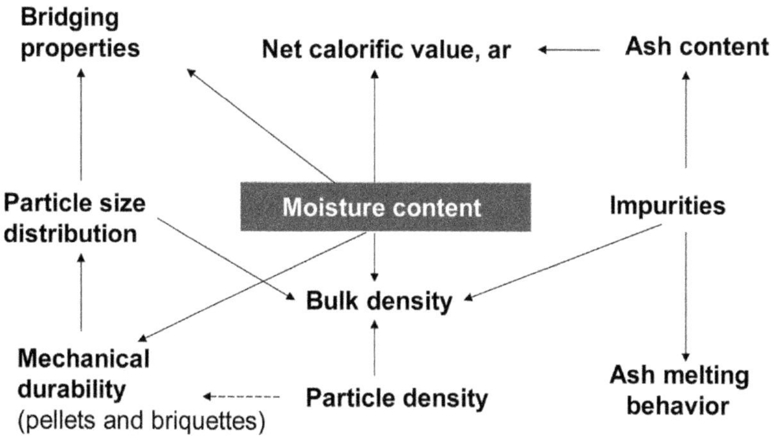

FIGURE 3.2 Interdependence of physical–mechanical biomass properties (Hartmann, 2007; Alakangas et al., 2016).

At the same time moisture content is a feature which can easily be manipulated, though often at an extra cost (see Chapter 4: Modeling Biomass Logistics). At harvest, moisture content will typically vary between 50% for wood and 80% for leafy herbaceous material. As moisture content is variable in the growing plant and also changes quickly after harvest, reported moisture content in databases should be seen as just an indication.

There are boiler technologies (1–10 MW_{th}) which can burn very wet woody biomass such as sawdust and fresh wood chips (50% moisture) with low emissions and high efficiency. In some cases a boiler can be equipped with flue gas condenser to recover energy from flue gases.

3.3.5.2 *Ash and Ash Melting*

Ash is the mass of inorganic residue remaining after combustion of a fuel under specified conditions, typically expressed as a percentage of the mass of dry matter in fuel. Ash content of biomass is generally defined as the remaining material after combustion at 550°C or 815°C. Ash consists mainly of silica and other oxides: SiO_2, Al_2O_3, Fe_2O_3, CaO, MgO, Na_2O, K_2O, TiO_2, P_2O_5, and SO_3. Ash is an important quality feature of biomass as explained earlier, because it generally confers a negative value. Ash has to be transported and needs to be disposed of, which causes extra costs.

Ash is undesirable in most conversion systems. A high ash content has drawbacks: (1) it does not contribute to energy production; (2) it may increase wear of the machinery; (3) it will generally cost money to discard ash; (4) thermal conversion systems are generally designed to a limited ash amount; (5) ash contains nutrients such as K and P, which have a value for plant growth and therefore for the forest or for agricultural land. These nutrients should preferably remain on the field or recycled.

Ash content for wood without bark is usually less than 0.5%. Bark and especially stumps can have higher ash content. Herbaceous biomass will have a higher ash content than wood, and it can vary between 2% and 20%. In grasses, leaves will generally have higher ash (and nutrient) contents than the pseudo stems. Separating leaves from pseudo stems has been proposed to increase quality.

The composition of ash differs according to plant tissue and plant type. But it is also influenced by soil type, available elements in the soil, fertilization used, and time of the year. For example, plants on a sandy soil will have less ash than plants on a clay soil (mainly due to higher silica uptake). Although plants only need some elements for plant growth, they only have limited capability for selective uptake of those elements, which are essential for their growth (Kirbey, 2012). Biomass can easily be contaminated by soil (on the outside of the plants) which can dramatically increase ash content of biomass samples, so called extraneous ash. It is important to be aware that many biomass ash data are at least partly determined by the external contamination after harvest. The composition of ash is also very relevant as this determines ash melting point.

The ash melting behavior is important because it is an indication of the risk of ash sintering and boiler fouling in thermal conversion technologies as explained earlier. Ash melting behavior for biomass is measured according to standard methods like the CEN/TS 15370-1:2006. Ashing temperature for biomass is typically 550°C. The following temperatures are usually measured: ash shrinkage starting temperature (SST), ash DT, ash flow temperature, ash hemisphere temperature in oxidizing conditions. These ash melting temperatures are determined by the ash composition (Alakangas et al., 2016). Having said that, it is still difficult to predict the exact value of the ash melting temperatures even when the ash composition is known. In general, high K and Na values decrease the ash melting temperatures while high Ca and Mg values increase the ash melting temperature. Indicators have been developed which predict ash melting behavior based on ash composition. Still, if different biomass types are mixed together, ash melting behavior cannot be predicted by separate biomass types, because ash melting behavior will change.

A new international standard (ISO 21404) for solid biomass is under development (planned in 2017) in which the ashing temperature is proposed to be 710°C ± 10°C. If ashing is done at 550°C, carbonates in Ca-rich solid biofuels, especially calcium carbonate, will mainly contribute to lowering the SST since decomposition of calcium carbonates occur from approximately 600°C. It may be more difficult to determine DT due to strong shrinkage of test pieces. If there is any doubt of incomplete ashing, then ashing shall be

FIGURE 3.3 Picture of agglomerated straw ash in a heating stove (by H.W. Elbersen).

continued at 710°C until complete ashing is obtained (Alakangas et al., 2016) (Fig. 3.3).

3.3.5.3 *Slagging and Fouling*

To predict the risk of slagging and fouling, some indicators have been developed which are useful if ash melting data are not available.

Burvall (1997) proposes a correlation between the Si/(Ca + K + Mg) ratio and the DT of ash in the context of reed canary grass. When the ratio is approximately 1, the DT is 1000–1100°C. If the ratio is 3, then DT is more than 1400°C (Alakangas et al., 2016). To predict the risk of slagging or fouling, Miles et al. (1996) developed the "alkali index":

$$\text{Alkali index} = (Na_2O + K_2O)/\text{Gross calorific value (HHV)}$$

A value below 0.17 indicates that slagging/fouling is very unlikely, while a value above 0.34 means that that slagging is very likely. An indicator for fouling risk is the "fouling index" developed by Teixeira et al. (2012). This index is based on the base-to-acid ratio of the ash using the formula:

$$\text{Fouling index} = ((Fe_2O_3 + CaO + MgO + K_2O + Na_2O)/(Al_2O_3 + SiO_2 + TiO_2))^*(K_2O + Na_2O)$$

A value below 0.6 indicates a low risk of fouling while a value above 0.40 indicates an extremely high risk of fouling.

3.3.5.4 *Chlorine (Cl)*

Chlorine is an essential micronutrient for higher plants. However, the average concentration in plant shoots for adequate growth is just 100 mg per kg of dry matter (0.01% of dry matter) (Kirbey, 2012). In practice the content can exceed 1% (of dry matter) in plants. High chlorine contents can be found if chlorine-containing fertilizers (KCl fertilizer) are used or when the biomass is produced on salt-containing soils. Reporting "typical" Cl contents for biomass types is therefore difficult. Chlorine content depends on crop management and soil type and hardly on the type of plant or plant variety.

Chlorine content together with Na and K content is important, especially for fluidized boiler combustion. Chlorine increases the chance of corrosion and harmful emissions in thermal combustion and is also unwanted in many chemical biomass treatments (second-generation ethanol production or pulp production). Fortunately, chlorine is easily flushed out of herbaceous biomass, when the biomass is weathered as is the case with winter-harvested reed, *Miscanthus*, switchgrass, and reed canary grass or old weathered (gray) straw. Weathering can reduce the chlorine content by 90%. The negative effect of chlorine in combustion can be partially undone if the S/Cl ratio is above 4, which is however, seldom the case in plants. For woody biomass higher Cl contents can be found in forest residues, which include needles or leaves (Alakangas et al., 2016).

3.3.5.5 Sulfur (S)

Sulfur is also an essential plant element that is needed for plant growth. The average concentration in plant shoots for adequate growth is 1000 mg per kg of dry matter (0.1% of dry matter). Plants will generally contain not much over this amount. Wood and similar material will have even lower contents. For woody biomass the S content is usually less than 0.05%. Sulfur can reduce the corroding effects of chlorine in combustion. For combustion a higher S/Cl ratio is positive, preferably above 4, which will seldom occur though. Therefore, Cl% is often used as an indicator for corrosion. S/Cl is mainly useful when someone is talking about coal or peat co-combustion.

3.3.5.6 Potassium (K)

Potassium is an essential plant mineral found in high amounts in living plant tissue. The average concentration in plant shoots for adequate growth is 10,000 mg per kg of dry matter (1% of dry matter). The variability of potassium in living tissue is not very high. A potassium content of 2% or higher is not often found. The potassium content may be much lower than 1% in dead and weathered tissues such as wood and winter-harvested reed, *Miscanthus*, switchgrass, *A. donax*, and reed canary grass or old (gray) straw. The potassium content of leafy biomass can have a value of €5–10 per ton of dry matter, which is especially relevant when using crop residues. Potassium is easily leached out of the tissue. Potassium is the main mineral causing sintering and boiler fouling in thermal conversion. Potassium content can be high in industrial wood residues including, for example, glue.

3.3.5.7 Sodium (Na)

Sodium is a nonessential element in most plants (Kirbey, 2012). There is therefore in general no minimum amount needed in plants. Sodium behaves similarly to potassium and it can replace the function of potassium in plants to some extent. In practice, plants can have very variable amounts of sodium in their tissues (as with chlorine). Contents can vary from close to 0% to more than 1%. The reported sodium contents in the database are therefore only indicative and depend more on the soil sodium content than on the type of plant. This is because the sodium content is strongly related to the amount of sodium in the soil. It is also easily leached out of plant tissues and it also increases the risk of sintering and boiler fouling in thermal conversion. Sodium content can be high in industrial wood residues, including, for example, glue. It is important to know the reactive K + Na content, especially for fluidized bed combustion (Alakangas et al., 2016).

3.3.5.8 Calcium (Ca)

Calcium is an essential plant element. The average concentration in plant shoots for adequate growth is 5000 mg per kg of dry matter (0.5% of dry matter). In living plant tissue, the variability of calcium is lower than for sodium. Calcium is not easily removed from plant tissues (unlike potassium or sodium). Calcium is generally positively correlated (increased ash melting temperature) to the ash melting behavior.

3.3.5.9 Magnesium (Mg)

Magnesium is also an essential plant element. The average concentration in plant shoots for adequate growth is 2000 mg per kg of dry matter (0.2% of dry matter). In living plant tissue, the variability of magnesium is lower than the variability of sodium. Just like calcium, magnesium is also not easily removed from plant tissues (unlike potassium or sodium). Magnesium is generally positively correlated to the ash melting behavior.

3.3.5.10 Silicon (Si)

Silicon is not an important plant element for most plants. Therefore, no minimum silica content of plants can be given. All plants grown in soil will contain some silicon in their tissues. However, the silicon concentration in the shoots varies considerably among plant species, ranging from 1 to 100 mg silicon per g dry mass (Broadly et al., 2012). Species which deposit a large quantity of silica in aerial parts have been termed silica-accumulator plants (Takahashi and Miyake, 1977). These may act as barriers of defense or add to strength or rigidity of plants (see Fig. 3.4 for an example).

Rice for example accumulates more than 10% silica in the leaves. Silica content is also related to transpiration and will mostly accumulate in the

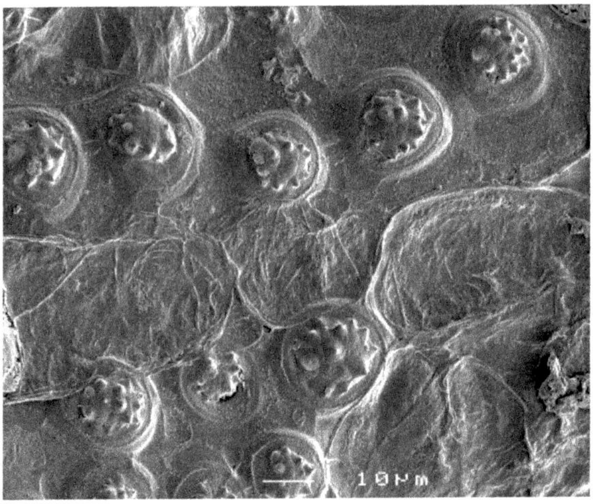

FIGURE 3.4 Scanning electron microscopy pictures of fresh empty fruit bunch fibrous strands showing the silica bodies (by Paulien Harmsen).

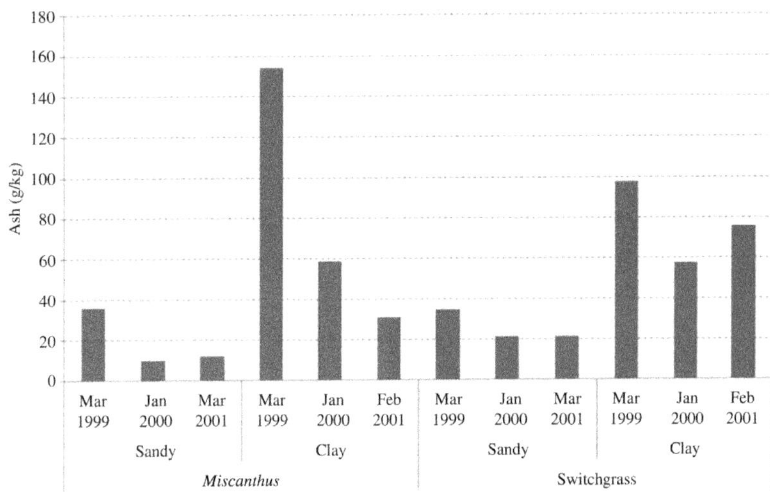

FIGURE 3.5 The ash content of *Miscanthus* and switchgrass grown on a sandy and on a clay soil harvested in winter or early spring illustrating the effect of soil type on ash content.

leaves. It is also observed that in grasses, ash content (mostly silica) is higher in clay than in sandy soil (Fig. 3.5).

Silica (SiO_2) is generally the most important element contributing to ash content of plants.

3.3.5.11 Nitrogen (N)

Nitrogen is the most important essential nutrient in plants. The average concentration in plant shoots for adequate growth is 15,000 mg per kg of dry matter (1.5% of dry matter). The nitrogen content can be as high as 5% in many tissues. If this is the case the value for other purposes than nonfood uses are likely to fetch a higher price. Nitrogen, which is mostly found in proteins, is generally high in seeds and young tissue. In older and mature tissue, nitrogen levels can be reduced considerably, mainly due to active translocation.

The nitrogen content of woody biomass is usually around 0.2%. Industrial wood residues including glue can have high N contents. Nitrogen does not flush out easily like chlorine or potassium. In thermal conversion, nitrogen is less of a problem than chlorine or potassium, but it may increase the risk of NO_x emissions, requiring costly emission controls. Emission reduction measures make a conversion technology more expensive and less economical to apply on a small scale (below 1 MW_{th}). Therefore the nitrogen content is indeed relevant to take into account, especially for smaller conversion systems.

3.3.5.12 Phosphorus (P)

Phosphorus is an essential element for plants. The average concentration in plant shoots for adequate growth is 2000 mg per kg of dry matter (0.2% of dry matter). In general, the phosphorus content will not be much above this. Phosphorus is not easily removed from plants, though content of phosphorus can be reduced in maturing tissue. The effect of phosphorus is less pronounced than for chlorine or potassium. Phosphorus does however have a cost and is becoming a scarce resource. Combustion can make phosphorus unavailable for recycling.

3.3.6 Energy Content of Biomass (LHV and HHV)

The energy content of biomass is expressed in net calorific (or lower heating value, LHV) or gross calorific (or higher heating value, HHV). In LHV the condensation heat of water is excluded. The HHV will therefore always be higher than the LHV especially, if the biomass contains water. The HHV (MJ/kg) of biomass is measured by a weighed portion of the analysis sample of the solid biofuel and is burned in high-pressure oxygen in a bomb calorimeter under specified conditions. The HHV is calculated from the corrected temperature rise and the effective heat capacity of the calorimeter, with allowances made for contributions from ignition energy, combustion of the fuel(s), and for thermal effects from side reactions such as the formation of nitric acid. Furthermore, a correction is applied to account for the difference in energy between the aqueous sulfuric acid formed in the bomb reaction and gaseous sulfur dioxide, i.e., the required reaction product of sulfur in the solid biofuel (EN ISO 17225-1:2014; Alakangas et al., 2016):

The net calorific value at a constant pressure for a dry sample (dry basis, in dry matter) is derived from the corresponding gross calorific value at a constant volume according to Eq. (3.1).

$$q_{\mathrm{p,net,d}} = q_{V,\mathrm{gr,d}} - 212.2 \times w(\mathrm{H})_{\mathrm{d}} - 0.8 \times [w(\mathrm{O})_{\mathrm{d}} + w(\mathrm{N})_{\mathrm{d}}] \qquad (3.1)$$

where

$q_{p,\mathrm{net,\,d}}$ is the net calorific value for dry matter at a constant pressure in joules per gram (J/g) or kilojoules per kilogram (kJ/kg);
$q_{V,\mathrm{gr,\,d}}$ is the gross calorific value for dry matter at constant volume in joules per gram (J/g) or kilojoules per kilogram (kJ/kg);
$w(H)_{\mathrm{d}}$ is the hydrogen content, in percentage by mass, of the moisture-free (dry) biofuel (including the hydrogen from the water of hydration of the mineral matter as well as the hydrogen in the biofuel substance);
$w(O)_{\mathrm{d}}$ is the oxygen content, in percentage by mass, of the moisture-free biofuel;
$w(N)_{\mathrm{d}}$ is the nitrogen content, in percentage by mass, of the moisture-free biofuel.

For the calculation of the net calorific value as received using Eq. (3.2), the result from Eq. (3.1) in joules per gram (J/g) or kilojoules per kilogram (kJ/kg) shall be divided by 1000 to get the result in mega joules per kilogram (MJ/kg).

Note: $\left[w(O)_{\mathrm{d}} + w(N)_{\mathrm{d}}\right]$ can be derived by subtracting from 100 (wt%) the percentages of ash, carbon, hydrogen, and sulfur.

3.3.7 The Net Calorific Value as Received

The net calorific value (at constant pressure) on as received (the moist biofuel) can be calculated on the net calorific value of the dry basis according to Eq. (3.2).

$$q_{\mathrm{p,net,ar}} = q_{\mathrm{p,net,d}} \times \left(\frac{100 - M_{\mathrm{ar}}}{100}\right) - 0.02443 \times M_{\mathrm{ar}} \qquad (3.2)$$

where

$q_{\mathrm{p,net,ar}}$ is the net calorific value (at constant pressure) as received in mega joules per kilogram (MJ/kg);
$q_{\mathrm{p,net,d}}$ is the net calorific value (at constant pressure) in dry matter in mega joules per kilogram (MJ/kg);
M_{ar} is the moisture content as received (wt%);
0.02443 is the correction factor of the enthalpy of vaporization (constant pressure) for water (moisture) at 25°C (in mega joules per kilogram (MJ/kg) per 1 wt% of moisture).

The energy content of biomass is therefore mostly dependent on net calorific value on dry basis and moisture. High ash content decreases net calorific value on dry basis. The dry and ash-free energy content of different biomass types are generally only marginally different and mostly determined by the lignin and oil content, which have a higher energy content than cellulose/hemicellulose and starch. Net calorific value as received (on wet basis) is calculated by Eq. (3.2).

3.3.8 Bulk Density and Particle Size Distribution of Biomass

The bulk density of biomass is expressed as kg/m^3 as received. It is a function of plant characteristics and harvesting, collection, and size reduction methods. It is an important characteristic, because it determines to a large extent the ease of handling and the cost of handling and transport (see Chapter 4: Modeling Biomass Logistics). Bulk density for wood chips and hog fuel is measured in a 50 L container according to EN ISO 17828:2016 (Alakangas et al., 2016).

Particle size distribution is an important property determining flowability and conversion behavior in thermal conversion. For example, gasification requires a more homogenous particle size distribution of biomass than grate combustion. Particle size distribution is analyzed by screening methods. For small combustion of wood pellets, it is important to know the amount of fine particles (<3.15 mm), because fine particles cause problems in storage and combustion.

3.4 METHODOLOGY FOR MATCHING BIOMASS FEEDSTOCKS WITH CONVERSION TECHNOLOGIES

3.4.1 Introduction

In the S2Biom project a biomass and technology matching tool was created that builds on the methodology that was described above, called Bio2Match. The tool was built using a database of conversion technologies and a database of biomass properties and thus brings together a large collection of data on both biomass properties and technology characteristics, in an interactive way. It should be able to help stakeholders in the bio-based economy to gain insight in bio-based value chains, by finding suitable matches. The tool is accessible online, via the S2Biom website (S2Biom, 2017).

3.4.2 The Bio2Match Methodology

The methodology for the Bio2Match tool was defined on the basis of the classification system described earlier, with fundamental characteristics (which cannot easily be modified) and physical characteristics (which can

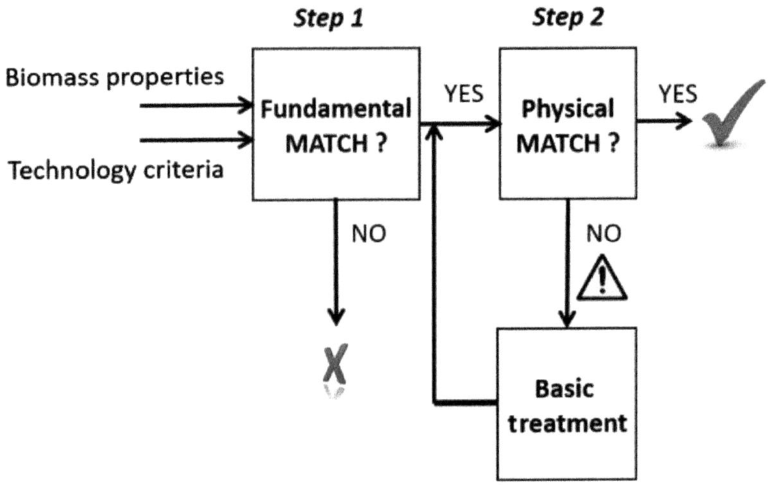

FIGURE 3.6 The Bio2Match tool methodology.

easily be modified). The procedure that the tool utilizes for matching each biomass and each technology is schematically shown in Fig. 3.6.

Depending on which type of technology is chosen (thermal, (bio-)chemical, or anaerobic fermentation), the relevant fundamental properties of the biomass are first compared with the technology criteria (step 1). When each biomass property class has an equal or lower number than the technology criteria for those properties, there is a fundamental match, and the tool subsequently investigates the physical properties (step 2). When the values for the main physical properties also match, the tool generates the answer "there is a physical match," indicated by a green (dark gray in print versions) traffic light symbol. When there is a fundamental match but no physical match, the tool generates the answer "there could be a match, if the biomass receives basic treatment," indicated by a yellow (light gray in print versions) exclamation mark (Fig. 3.7). When there is no fundamental match, the tool does not proceed to step 2, but generates the answer "there is no match," indicated by a red (black in print versions) traffic light symbol.

3.4.3 Bio2Match in Practice

The user interface of Bio2Match is shown in Fig. 3.7. The left two columns are for the selection of biomass types and technologies. The user can select any number of biomass types and conversion technologies that he or she is interested in. The center screen in Fig. 3.7 shows which technologies match to which types of biomass, based on the methodology that was described earlier. A green (dark gray in print versions) check mark indicates a match, while a yellow (light gray in print versions) exclamation mark indicates a

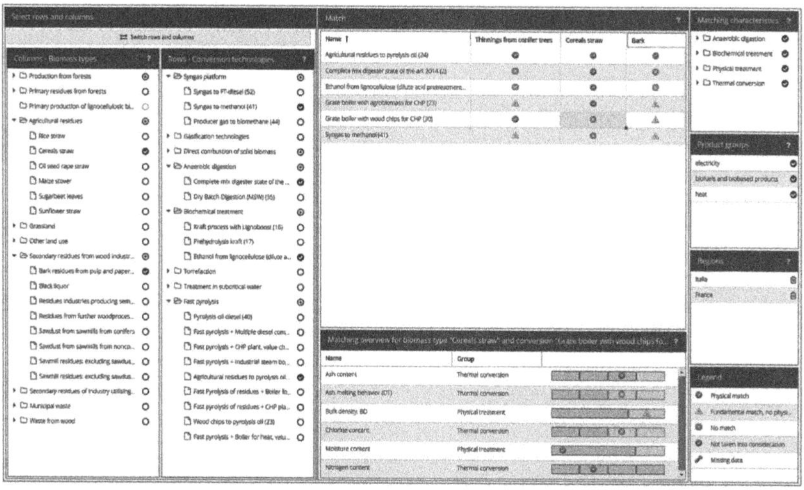

FIGURE 3.7 Screenshot overview of Bio2Match tool showing matching results for a selection of biomass types with conversion technologies.

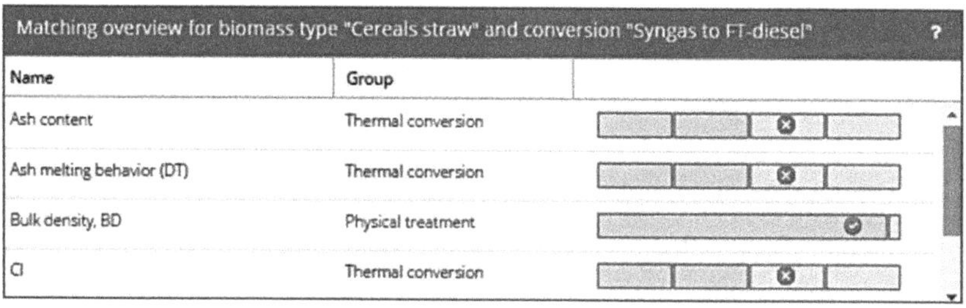

FIGURE 3.8 Matching overview pane in Bio2Match tool.

fundamental but no physical match, and a red (black in print versions) cross indicates that there is no fundamental match (see Fig. 3.7 lower left panel "Legend").

The user can then select a single biomass–technology combination to find out in more detail why a feedstock does or does not match to a certain conversion technology. The matching classification system is visualized in the screen below the center screen with four blocks per property, in which green (dark gray in print versions) blocks represent the biomass quality that the technology can handle (Fig. 3.8). When, for example, only the left block (class 1) is green (dark gray in print versions), it means the technology can handle only feedstock of class 1 quality, when, for example, all four blocks are green (dark gray in print versions) the technology can handle all biomass quality classes. The matching symbol (green (dark gray in print versions) ok sign or red (black in print versions) cross) within one of the blocks represents the actual quality of the selected biomass. If the symbol is positioned in a green (dark gray in print versions) block there is a match and the label turns green (dark gray in print versions), if it is situated in a red (black in print versions) block that means the technology cannot handle that feedstock quality and the label turns red (black in print versions) in the case of a fundamental property or yellow (light gray in print versions) in the case of a physical property (see Fig. 3.8).

In this way the user can quickly identify which biomass property is responsible for a mismatch between a biomass type and a technology type. In the example of a grate boiler that is designed for the combustion of wood chips, cereal straw does not match as a feedstock, as depicted in Fig. 3.7 (the highlighted biomass–technology combination). It can be seen in the screen that is at the center-bottom of Fig. 3.7 that the chlorine content and the ash melting temperature are responsible for the fact that there is no match between this type of biomass and this conversion technology.

It is important to note that the matching information is based on typical values for biomass types, while in practice the properties of biomass depend heavily on the growing and harvesting conditions (soil, fertilizer, season of harvest, etc.). Therefore in the Bio2Match tool users can also choose to display the best and worst-case values of the biomass properties. This can be done by clicking on the overhead bar of the matching overview pane and then selecting additional columns, as shown in Fig. 3.9.

Another outcome could be the case of a match, in which a user may find in the matching overview screen that a certain biomass type is of such high quality that it would be less than optimal to use it for a certain technology, because another biomass type of lower quality could also be converted by that same technology. Such findings may help to optimize the use of biomass in a region.

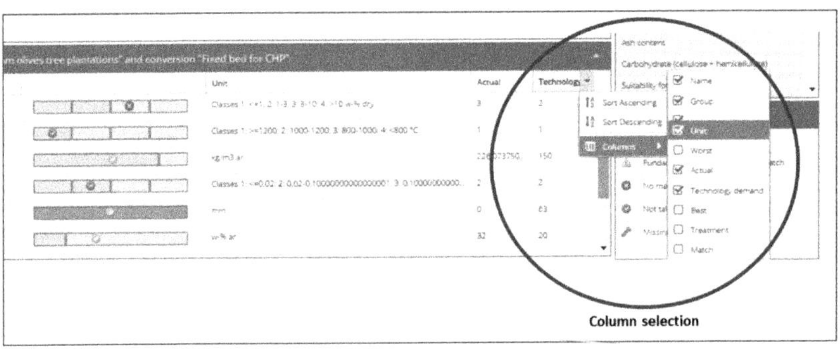

FIGURE 3.9 Screenshot of Bio2Match tool pane to obtain information on best and worst-case values of biomass properties for the selected technology.

With Bio2Match the S2Biom project aims to provide support for the development of strategies for best ways to realize a bio-based economy in Europe. The tool is intended to be user-friendly to biomass and/or technology experts, but specifically also to other stakeholders in the bio-based economy, such as for example policy makers and entrepreneurs. Bio2Match brings together a large collection of data on both biomass properties and technology characteristics, in an interactive way. It should be able to help stakeholders in the bio-based economy to gain insight in bio-based value chains.

REFERENCES

Alakangas, E., Hurskainen, M., Laatikainen-Luntama, J., Korhonen, J., 2016. Properties of indigenous fuels in Finland, VTT Technology 272, 222 p. + app, 23 pp. <http://www.vtt.fi/inf/pdf/technology/2016/T272.pdf>.

Al Seadi, T., Rutz, D., Prassl, H., Köttner, M., Finsterwalder, T., Volk, S., et al., 2008. Biogas Handbook. ISBN 978-87-992962-0-0.

Broadly, M., Brown, P., Cakmak, I., Rengel, Z., Zhao, F., 2012. Beneficial elements. In: M., Petra (Ed.), Marschner's Mineral Nutrition of Higher Plants, third ed. Elsevier, San Diego, CA.

Burvall, J., 1997. Influence of harvesting time and soil type on fuel quality in reed canary grass (*Phalaris arundinacea* L.). Biomass and Bioenergy Vol. 12 (No 3), 149–154.

Hartmann, H., 2007. Physical–mechanical fuel properties—significance and standard determination methods. Workshop: pre-normative research on solid biomass fuels for improved European standards, Athens, Greece, 15 pp.

Kirbey, E., 2012. Introduction, definition and classification of nutrients. In: M., Petra (Ed.), Marschner's Mineral Nutrition of Higher Plants, third ed. Academic Press, Elsevier, San Diego, CA.

Miles, T.R., et al., 1996. Alkali Deposits Found in Biomass Power Plants. A Preliminary Investigation of Their Extent and Nature. National Renewable Energy Laboratory, Golden, CO, Report NREL/TP-433-8142.

S2Biom project website, 2017. <http://www.S2Biom.eu> (last accessed March 2017).

Takahashi, E., Miyake, Y., 1977. Silica and plant growth. Proceedings of the International Semin. Soil Environ. Fert. Manage. Intensive Agric. 603–611.

Teixeira, P., Lopes, H., Gulyurtlu, I., Lapa, N., Abelha, P., 2012. Evaluation of 529 slagging and fouling tendency during biomass co-firing with coal in a fluidized 530 bed. Biomass Bioenergy 39, 192–203.

FURTHER READING

Elbersen, W., Bakker, R., Harmsen, P., Vis, M., Alakangas, E., 2015. A selection method to match biomass types with the best conversion technologies. S2Biom Project Grant Agreement no. 608622 S2BIOM Deliverable D2.2.

Elbersen, W., Alakangas, E., Elbersen, B., Annevelink, E., Ramirez Almeyda, J., Lammens, T., 2016. D2.4 explanatory note accompanying the database for standardized biomass characterization (and minimal biomass quality requirement for each biomass conversion technology). S2Biom Project Grant Agreement no. 608622.

Lammens, T., Vis, M., de Groot, H., Vanmeulebrouk, B., Staritsky, I., Elbersen, B., et al. 2016. BIO2MATCH: a tool for optimizing the match between lignocellulosic biomass and conversion technologies. Proceedings of the 24th European Biomass Conference and Exhibition, Amsterdam.

Lammens, T., Vis, M., te Raa, R., van den Berg, D., Kärki, J., Uslu, A., et al. 2016. D2.3 database of biomass conversion technologies S2Biom Project Grant Agreement no. 608622.

Vis, M., te Raa, R., van den Berg, D., Knoef, H., Alakangas, E., Karki, J., et al., 2015. D2.1: a method for standardized biomass characterization and minimal biomass quality requirements for each biomass conversion technology. S2Biom Project Grant Agreement no. 608622.

Woodenergy, 2014. Wood as a fuel. <www.woodenergy.ie/woodasafuel>.

Chapter 4

Modeling Biomass Logistics

Bert Annevelink, Perttu Anttila, Kari Väätäinen, Benoît Gabrielle, Daniel García-Galindo, Sylvain Leduc and Igor Staritsky

4.1 INTRODUCTION

4.1.1 Biomass Logistics

The efficiency at which biomass can be used for producing bioenergy and bio-based products is a key issue when designing biomass supply chains. In this respect biomass presents a number of logistical challenges as its quality and handling characteristics often restrict options for efficient logistics and respective efficient conversion. The logistics of a biomass supply chain may include several logistical components (Fig. 4.1) such as handling, pretreatment, storage, and transport. Pretreatment technologies like comminution (size reduction), compaction/densification, and drying are desirable in the biomass supply chain to convert the biomass "as received" at the roadside (an amount in t, with a given cost €/t) to an intermediate biomass product with the required quality at the gate of the biomass conversion facility (an amount in t, with certain costs €/t at the factory gate). Storage bridges gaps in time between supply and demand. Finally, transport is required to collect the biomass from many different sites of origin to a specific location "at the gate" of a certain conversion plant. A biomass value chain can be split up into three separate parts:

1. The biomass production, including harvesting
2. The logistical chain with several logistical components
3. The conversion process.

4.1.2 Main Challenges

Several factors affect biomass quality for thermal and biochemical conversion and design of a supply chain. The issues around quality have been described thoroughly in Chapter 3 of this book. All need to be addressed through an optimal design of sustainable biomass supply chains. Following is a summary of these factors.

Modeling and Optimization of Biomass Supply Chains.
DOI: http://dx.doi.org/10.1016/B978-0-12-812303-4.00004-5

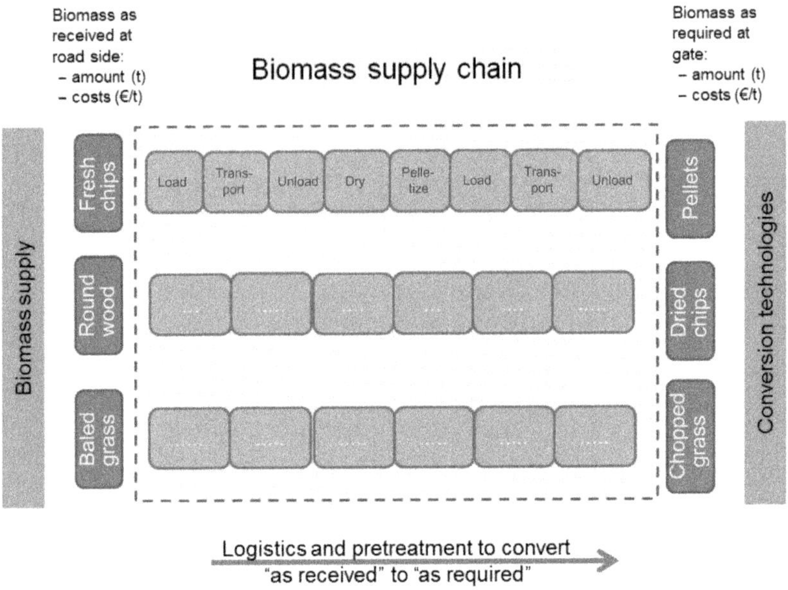

FIGURE 4.1 Schematic representation of a biomass supply chain.

1. Ash content and composition—Many biomass types contain larger amounts of inorganic components (generally referred to as ashes) compared to the clean wood fuels that are currently being mobilized and used for (co-)firing. In addition, the composition of the ashes is such that the biomass fuels may exhibit a poor quality for thermal combustion or gasification, as they lead to a relatively low ash melting point. This in turn leads to ash slagging and agglomeration problems in both, large and small, thermal conversion installations. Furthermore, if chlorine is present in biomass, enhanced boiler tube corrosion may occur. These ash-related problems are one of the main reasons that ash-rich biomass types (e.g., straw from annual crops) are often not used for energy conversion.
2. Heterogeneity across biomass types—Many available biomass resources are currently not used for energy conversion, as they are very heterogeneous in nature, which makes the large-scale conversion in energy expensive. For instance, in many large boiler systems, it is not possible to accept biomass fuels with a wide variety in shape and size of particles, as it makes blending with coal fuels and/or firing in fluidized bed systems problematic. Moreover, the moisture content (MC) of biomass can vary a lot affected by the time and type of storing and weather conditions. MC has an impact on transport efficiency, energy content of biomass,

requirement of premixing, storing, drying, burning process, desirability of biomass for certain conversion processes, and finally supply costs. Thus, heterogeneous biomass types are either not used for conversion, or they need to undergo costly pretreatment, such as size reduction (milling), screening, and/or pelleting.

3. Spatially scattered biomass supply chains—In combination with the previous two quality aspects, the spatially dispersed availability of many biomass types often hampers their conversion to bioenergy and bio-based products, and increases costs for collection and transportation.
4. With increasing demand transport distances will increase—Therefore, industries are seeking cost-effective alternatives to the traditional direct truck transport. Logistical concepts with multimodality transport might provide cost-effective solutions for long transport distances.
5. Seasonal variation—The factor most affecting the profitability of the entrepreneurs working with biomass harvesting, chipping, and transport is seasonal variation of biomass supply. The numbers of machines and workers have to be scaled according to high season, which leads to low utilization rate in low season. Biomass yards/terminals could be used to alleviate the effects of seasonal variation.

4.2 LOGISTICAL COMPONENTS

A logistical component is one of the links to solve mismatch problems in the biomass value chain from biomass production to final conversion (Annevelink et al., 2014, 2015a). Examples are pretreatment, storage, and transport technologies that are needed to deliver biomass of a specified quality at the correct moment to a processing technology. Logistical components can be divided into subcategories related to their technical properties, biomass input requirements, and capacity range. The quality of the biomass is modified in the value chain so that in the end its form is more suitable for the conversion process. Logistical challenges can take different forms leading to the following main categories of logistical components:

- Biomass pieces that are too large or too irregularly shaped (inhomogeneous quality) → comminution (size reduction)
- Biomass density is too low → compaction/densification
- Biomass is too wet (relatively high MC) → drying/fuel mixing
- Not in place at the correct logistical component or process → biomass handling
- Biomass still growing on the field or distributed in space → harvesting/collection

- Biomass is contaminated with soil particles, etc. → sieving, screening, washing (other pretreatments)
- Biomass is not available in each period of the year (seasonal supply patterns) → storage, densification
- Biomass is scattered over many sources locations) → transport technologies.

4.2.1 Comminution (Size Reduction)

In most cases, and especially with loose biomass, comminution (size reduction) diminishes the volume of the biomass material and thus increases the bulk density. Follow-up processes like handling and transport will be facilitated by size reduction, especially for heterogeneous, loose materials. Furthermore, the physical properties of the biomass will be adjusted, e.g., the particle size distribution will be modified. A method often used to reduce the size of biomass is chipping. Many different types of chippers are being used (disk, drum, and screw chippers), and each will have a specific influence on the quality of the produced biomass chips. If fresh chips need to be stored for a longer period, they will need to be dried first to avoid problems with biological activity (see Section 4.2.3). An alternative way is to chunk the biomass, which will produce larger particles that are easier to dry and that can therefore be stored longer. Shredding is another possibility, being similar to chipping, but producing materials with, in general, more variable particle size, a higher number of fines, and without a clean cut.

4.2.2 Compaction/Densification

Compaction/densification can be used to produce a uniform material with favorable physical and mechanical properties. Densification methods for lignocellulosic biomass include baling, briquetting, and pelletizing. Biomass is densified under high pressure into briquettes (diameter 50–100 mm, length 60–150 mm) or pellets (6–12 mm). Sometimes this requires a binding agent (other than the lignin already present in the biomass). The economy of pelletizing strongly depends on the volume and the costs of the raw material. An advantage of pellets is that they are dry, clean, and mechanically stable with high energy content. Furthermore, pellets can flow more easily, which facilitates loading and unloading and wood pellets, for example, are standardized in Europe (EN 14961-2). So far pellets are used in the bioenergy markets: both industrial (co-firing in coal fed power plants) and household markets (stoves). Baling is a densification process of special interest for herbaceous materials. However, recently balers have also been developed to compress residues like woody energy crops or tree branches.

4.2.3 Drying

Biomass can be dried to reduce the MC and decrease dry matter loss (DML) while storing. This influences (1) efficiency of conversion processes, (2) storage options, and (3) transport costs.

For thermal conversion processes like combustion or gasification, dry biomass is needed. That way less energy content is lost due to evaporation. Moreover, different thermal conversion processes demand different MCs. Typical gasification processes require biomass with an MC in the range of 10%–15%. Combustion can also deal with higher MCs depending on the design of the boiler. In some case it is also possible to combust freshly harvested biomass, e.g., wood chips, with an MC of 40%–50%. In most cases drying is not necessary for anaerobic digestion processes. Other processes like HydroThermal Upgrading (HTU) specially require relatively wet biomass so they also do not need drying. Biomass that is too wet (more than 20% MC; Table 4.1) is

TABLE 4.1 Bulk Density and Moisture Content of Different Biomass Fuels (Alakangas and Virkkunen, 2007; Woodfuel Handbook, 2009; Woodenergy, 2014)

Name	Manifestation	Bulk Density Fresh (kg/loose m^3)	Moisture Content Fresh (wt%)
Whole tree	Chips	250–350	45–55
Stem wood	Chips	250–350	40–55
Stump	Chips	200–300	30–50
Logging residue	Chips	250–400	50–60
Log wood (oven-ready)	Logs	240–320	20–25
Wood residue	Chips	150–300	10–50
Wood	Pellets	550–650	7–8
Plywood	Chips	200–300	5–15
Sawmill residue	Dust	250–350	45–60
Cutter	Chips	80–120	5–15
Grinding residue	Dust	100–150	5–15
Straw	Chopped	80	12–20
Straw	Pellets	550–650	8–10
Miscanthus	Chopped	110–140	8–20

perishable. By drying wet biomass, it will be easier to store the material for a longer time without danger of heating, composting, ignition, or molds. Furthermore, transport of dried biomass can be more efficient because less water needs to be moved.

Several methods are available for drying biomass. The cheapest alternative is to dry passively in the open air. This could be done in several configurations, e.g., in covered piles on bearers (see more storage methods in Section 4.2.7). Usually outdoor drying can lead to a reduction of MC to about 20%–30%. Factors that influence drying biomass are initial MC, shape and size of the biomass particles, bulk density during storage, method of storage, and weather conditions (wind, temperature, rainfall, and relative humidity). When passive drying does not lead to the required results, one can move to active/forced drying. However, this demands energy input and additional costs for a drying installation. Sometimes it is possible to use heat surplus from other processes.

4.2.4 Biomass Handling

Biomass handling is required throughout the biomass value chain to move the biomass between different logistical components, e.g., loading biomass from storage to a means of transportation. Another use for biomass handling is to move biomass from one position to another, where the transport distance is relatively low.

4.2.5 Harvesting/Collection

Integrated methods have been developed for harvesting biomass in agriculture and forestry (both for the main product and for residues). A restrictive factor is the fact that purpose-grown biomass plants and primary agricultural and forestry residues are always produced de-centrally, distributed all over the supply region. Therefore, the biomass still needs to be transported to a more central location. Furthermore, biomass often can only be harvested in a certain time, which requires some storage capacity to ensure a year-round supply. The total amount of biomass that can be removed from an agricultural field or a forest plot/site is limited by various reasons, such as preventing erosion and soil compaction, maintaining soil fertility, and ecological reasons. Moreover, agricultural residues often have alternative uses (i.e., livestock applications in the case of straw) that can limit the potential for energy conversion. In other cases, such as for forestry residues, the resource accessibility and the high harvesting and transport costs can be a major drawback for the full deployment of the theoretically available resources. During harvesting or collecting, biomass pollution (e.g., with soil or stones) needs to be prevented. Collection is a process needed to obtain secondary and tertiary residues on the location where they originate. This can be both centrally (e.g., at a food processing industry) and de-centrally (e.g., at demolition activities).

4.2.6 Other Pretreatments That Influence Biomass Quality

This category covers a broad collection of other pretreatments that might occur within the biomass supply chain. Different subcategories of logistical components for other pretreatments are biological pretreatments (fungi), blending, conservation (e.g., silage), dewatering, separation, sieving, sorting out metal with a magnet, ultrasonic pretreatment, and washing.

Torrefaction, Torwash, and HTU could also be seen as pretreatment technologies. However, in this book, they are described as conversion processes in Chapter 3, Lignocellulosic Biomass Quality: Matching Characteristics With Biomass Conversion Requirements.

4.2.7 Storage

Storage is important in relation to buffering biomass either over a longer period somewhere in the supply chain (e.g., to compensate for seasonal effects of biomass supply from forests and nature areas, where it is impossible to harvest year-round) or on the short term just before delivering to the final conversion process. Both relatively dry biomass (e.g., wood) and wet biomass (e.g., grass) can be stored. During storage, since biomass is an organic material, whenever its MC is higher than 20%, fermentation may occur, leading to heat development. This natural process on the one hand helps to reduce the MC but on the other hand can also have a negative effect generating DML, energy loss, ignition risks, and health problems for the operators (related to microorganism growth).

When choosing a storage method it is also important to take into account handling: how can biomass be received efficiently and how can it easily and safely be taken out of the storage facility again in order to supply it to the next link in the value chain.

Storage problems can be reduced through: (1) minimizing storage duration, (2) correct management and control of the storage, (3) the use of flexible storage facilities that can handle a wide range of materials, (4) investing in adjustments that match the characteristics of the specific biomass to be stored, and (5) pretreating the biomass prior storage (elimination of fines, forced drying, etc.).

A bigger storage capacity at the end users side can lead to less frequent supply visits, lower biomass costs (e.g., due to discounts), and a bigger reserve (with less danger of a shutdown of the conversion process). However, a bigger storage is more difficult to be controlled and optimized, so it is more likely to suffer from the problems mentioned earlier.

The subcategories for storage are a result of the combination of several characteristics:

- Indoors vs outdoors
- Covered vs uncovered
- Base type: asphalt, bare soil, bearers, or concrete floor
- Permanent storage structure type: bunker, container, silo or tank
- Temporary bulk form type: big bag, ensiled, pile or stack.

Storage can also be needed in case of excessive quantities of biomass (e.g., natural disturbances). The large and sudden supply of raw material that subsequently becomes available may by far exceed the conversion and sales capabilities of the sawmilling industry (Stodafor, 2004).

4.2.8 Transportation Technologies

Biomass can be transported locally, regionally, and globally. However, it is difficult to specify the exact distances between these subareas, because this depends on factors like bulk density, means of transportation, and density of the infrastructure (road, rail and waterway networks). Roughly speaking local supply may involve distances ranging from 0 to 30/50 km, regional supply distances vary between 30/50 and 250 km, and global sourcing distances are larger than 250 km.

It might not be necessary to pretreat the biomass yet for local transport (<30 km) to a first depot (biomass yard). However, for transport over longer distances (> 250 km), biomass will usually need to be pretreated at a biomass yard near the biomass production/extraction point to increase its bulk density and make it storable. Pellets are an example of a biocommodity that is already transported worldwide in bulk carriers (Sanders et al., 2009).

Transport on short and medium distance will most likely be performed on the road. Many transport means are available for this, like farm trailers, bulk vans, removable cargo containers, tipper trailers, and walking floor trailers with a prime mover. For long (and sometimes medium) distances, other transport options like rail, inland waterways, and sea might also be possible. Transport types are then closed or open train wagons, wagons with the WoodTainer system, and several types of barges and bulk carriers. Multimodal logistical networks make use of a combinations of transport means, e.g., bulk carriers to transport pellets to a large port (like Rotterdam) and dry bulk cargo barges to further transport the pellets to an inland power plant (Kisslinger, 2004). Another example is a combination of road transport to a railway yard and then railway transport to a large biomass power plant in Sweden (Ljungblom, 2011).

4.3 LOGISTICAL CONCEPTS

A logistical concept is broader and more general than a specific biomass value chain (Annevelink et al., 2015b, 2016e). Often several possible biomass value chains fit within that general logistical concept. Examples of logistical concepts are:

- pretreatment (e.g., comminution and densification) integrated with harvesting/collecting vs stand-alone pretreatment later on in the biomass value chain;
- indirect supply to the final conversion location through biomass yards (often in combination with intermediate storage and pretreatment) vs direct supply from roadside;
- multimodal transport (combination of different transport types) vs only one transport modality (road, water, rail);
- European/worldwide biomass value chains based on standardized biocommodities (e.g., wood pellets, ethanol, or pyrolysis oil) vs regional biomass value chains based on locally sourced "raw" biomass.

4.3.1 Pretreatment Integrated With Harvesting/Collecting

In the agricultural sector, there is already a long tradition of integrating several operations during the harvesting process. A good example is the development of the combined harvester, where cutting the stalks with the grain and threshing it to separate the grain from the straw are combined in one machine. Sometimes this is even combined with the baling of the straw. Herbaceous lignocellulosic crops such as *Miscanthus*, triticale, or sorghum are usually harvested with conventional agricultural machinery having undergone some adaptations, and may involve up to three passes in the case of bales: one to cut stalks, another one to rake, and them and baling itself a few weeks afterward to let the stalks dry on the field. One of the concepts tested in the Logist'EC project (see Section 4.5.1 for a short description of the project) consisted of a single-pass system, which proved successful at reducing fuel and economic costs, while producing high-density bales that are easier to transport or store (Gabrielle et al., 2015).

In case of biomass harvest from agricultural prunings, the integrated collection concept differs from annual crops. Pruning and fruit/olive/grape harvesting is done in two different moments of the crop cycle, and so, the integration of main product harvesting with the harvest of the residue is not possible. Pruning is carried out usually manually with shears or with mechanically assisted tools (e.g., electric shears or chainsaws), even though there is a trend to execute nonselective mechanized prepruning operations. The prepruning takes away important parts of the shoots and branches, to reduce the manual work to be carried out later on by workers. Pruning collection can be integrated in next ways:

- Mechanized prepruning: By building new machinery able to convey the shredded pieces of wood into a collecting system (instead the default spread of wood pieces on the soil).

- Pickup and treatment: The current pruning biomass harvest consists of machinery with a pickup system feeding the branches into the shredder, chipper, or baling system. Harvest requires in many cases a previous windrowing operation. Therefore integration can be improved by including windrowers with the harvester, or by adapting the manual pruning work so that the pruning is placed in the center of rows.

Also in the forestry sector integration of chipping with felling is possible in easy terrain, if the soil-bearing capacity is good and yarding distances short. In Finland and Sweden, chipping in the forest is no more practiced due to logistical challenges and high costs (Routa et al., 2012). For new nonfood biomass supply chains, integration of certain pretreatments with harvesting or collecting might be advantageous as well. This can already be seen in the three recent EU projects EuroPruning (García-Galindo et al., 2016a,b), Infres (Alakangas et al., 2015), and Logist'EC (Gabrielle et al., 2015) dealing with new concepts to optimize biomass logistics (see Sections 4.5.1 and 4.5.2). Advantages and disadvantages of the various logistical concepts are given in Table 4.2.

4.3.2 Biomass Yards

Several types of inhomogeneous biomass supplied by different sectors (e.g., agriculture, forestry, nature management) can be collected efficiently on a central location (biomass yard) in a region. There the biomass can be pretreated into a standardized intermediate product (biocommodity) for further processing by industry. In some cases biomass is also directly converted into an end product at the biomass yard. The goal of a biomass yard is to produce

TABLE 4.2 Advantages and Disadvantages of Four Logistical Concepts

Advantages	Disadvantages
Pretreatment Integrated With Harvesting/Collecting	
• Save one or more extra rides with a machine on the field	• More expensive machine for initial investment
• Lower overall costs per tonne for the combination of operations	• More complex machinery including system with multiple purposes
• Better biomass quality (e.g., less contaminated with soil)	• Heavier machine which needs to be compensated (with larger tires) to avoid soil compaction
• Save extra biomass handling	• Slower harvesting rate
• Possibility of direct loading in transport vehicle	• Failure in the biomass system may abort the harvest of the main product

(Continued)

TABLE 4.2 (Continued)

Advantages	Disadvantages
Biomass Yards	
• Production of biocommodities (intermediate products) on specification with a guaranteed quality to supply different buyers	• Insufficient experience with license to operate new type of biomass yards (not only woody biomass)
• Cheaper alternative for agricultural and horticultural companies to get rid of their biomass(residues)	• Success is only possible when the leading partner is clear
• Different types of biomass can be combined, and thus treated more efficiently with the correct technology at a central location	• Biomass is an infant industry: only a limited number of buyers and limited price building and competition
• Knowledge of existing organic residues collecting companies can be easily applied and integrated	• Inserting a biomass yard in the existing logistical biomass chain requires extra investment and operational costs
• Alleviation of the effects of seasonal variation	• "Theoretical" biomass yard concept still has to prove itself (partially)
	• Biomass streams ideally need to remain separated from each other to enable further processing
Multimodality Transport	
• Enables optimal choice of transport type for each section of the transport route	• More handling during transhipments
• For longer transport distances cheaper transport modes (ship or train) can be chosen	• Not all locations can be reached by alternative transport modalities
	• Higher loading and unloading costs
Biocommodities	
• Standardization	• Sufficient market volume is needed
• Easier to trade (inter)nationally	• Higher pretreatment costs need to be compensated (e.g., by lower transportation costs)
• Better transportability	
• Increased storability through stability	
• Better quality that can be measured	

a homogeneous output stream with the required specifications that can be traded as a biocommodity (Sanders et al., 2009). A biomass yard has (1) technical–operational tasks and (2) management (of supply, fuel quality) and trade tasks. Concepts that resemble a biomass yard are mentioned in several European countries like Austria, Germany, Sweden, and Finland. The concept is also mentioned in the United States and Canada. A biomass yard can also play a role in the process of guaranteeing the year-round availability of biomass. The biomass will be stored at the biomass yard until the exact moment when it is needed.

4.3.3 Multimodality Transport

The traditional method for transporting biomass is road transport. However, alternatives could be transport by rail or by waterway. The choice for a certain type of transport is determined by: (1) transport distance and (2) accessibility. The feasibility of multimodality transport increases in combination with the use of a biomass yard and/or with pretreatment. The average truck transportation distance of timber in Finland in 2014 was 107 km (Strandström, 2015). For a train transportation chain the corresponding distance was 322 km including 50 km pretransport by truck and for a waterway transportation chain 295 km. Energy wood is predominantly transported by trucks. If the transportation distance is less than approximately 50–70 km, the cheapest transport mode is uncomminuted by a truck (Tahvanainen and Anttila, 2011). With longer distances transport of comminuted material with a chip truck is more economical. Train transport of energy wood is currently not economically viable in Finland (Nivala et al., 2015).

4.3.4 Biocommodities

A biocommodity is a standardized form of biomass. The need for the development of biocommodities was described by Sanders et al. (2009). The main reasons for developing biocommodities are security of supply, the need for quality standards and quality control, and facilitation of trade. Possible examples of biocommodities that will be further developed are wood pellets, pyrolysis oil, torrefaction pellets, and biosyngas. It is not completely clear yet which biocommodities will arise, and this will certainly be a stepwise introduction. However, biocommodities will certainly have a number of properties that include transportability, stability, sufficient market volume, year-round availability, competitive costs, easily standardized in uniform quality characteristics for specific applications, and easy and quick quality measurements possible. Biocommodities could be produced, for example, at a biomass yard and they will greatly influence the setup of a biomass value chain. They will enable transportation of the biomass over a much longer distance, thus facilitating international trade.

4.4 AN OVERVIEW OF SOME LOGISTICAL TOOLS

Three logistical assessment methods have been used in the S2Biom project (Annevelink et al., 2016a,d):

1. BeWhere for the European and national level
2. LocaGIStics for the Burgundy and Spanish case on the regional level
3. Truck Transport Logistics simulation model for the Finnish case.

BeWhere and LocaGIStics are closely interlinked (Fig. 4.2) so that LocaGIStics can further refine and detail the outcomes of the BeWhere model, and that the BeWhere model can use the outcome of the LocaGIStics model to modify their calculations if needed. BeWhere, an optimization model, optimizes the plant location, and LocaGIStics, a simulation model, simulates the collection points for a given plant provided by the BeWhere model. Table 4.3 highlights the main differences between BeWhere and LocaGIStics.

Witness simulation software (Lanner, 2016) was used for building the simulation environment for the Truck Transport Logistics of timber in the Finnish case. The simulation is based on a discrete-event approach, where the system is changing as discrete events in predetermined timespans. Discrete-event simulation is a proper method for modeling complex environments, which have a lot of interactions between the modeled objects, where stochasticity is included in the system and where system operations are unstable and time dependent. Timber truck transport logistics indeed have this kind of complexity.

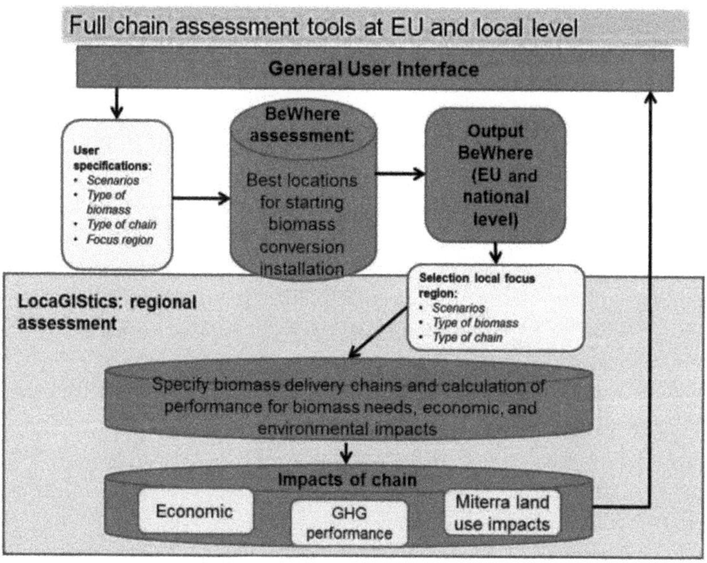

FIGURE 4.2 Relation between BeWhere and LocaGIStics.

TABLE 4.3 Comparison Between Functionality of the Logistical Tools BeWhere, LocaGIStics, and Truck Transport Logistics

BeWhere	LocaGIStics	Truck Transport Logistics
• Supply chain optimization	• Supply chain simulation	• Supply chain simulation
• National level	• Regional level	• Regional level
• Policy maker	• Project developer	• Transport company
• Rough grid biomass availability map	• Finer grid biomass availability map	• Biomass situated in storage points
• Determine the optimal geographic location of production plants	• Use one of the plant locations optimized from BeWhere and refine it	• Optimize collection process of biomass

4.4.1 BeWhere

The BeWhere model (www.iiasa.ac.at/bewhere) supports the development of EU-wide and national strategies to develop an optimal network of biomass delivery chains (Leduc, 2009; Leduc et al., 2012; Wetterlund et al., 2013; Natarajan et al., 2011). The basis of this tool is a techno-economic spatial model that enables the optimal design and allocation of biomass supply chains (at national level) based on the minimization of the cost and emissions of the full supply chain and taking into account economies of scale, in order to meet a certain demand. It considers input on biomass cost–supply, conversion technology specifications and logistical and pretreatment technologies and demand categories (for different scenarios). It takes into account the already existing production plants and local energy demand (provided the information is included in the tool). BeWhere provides as output a network of existing and suggestions for new biomass conversion and pretreatment chains according to optimal selection of technology, their location and capacity, the costs of each segment of the supply chain, the total bioenergy and biomaterial demand (depending on which technologies can be feasibly included in the tool), and avoided emissions at different geographic levels (regional, national, and European level). The spatial resolution in BeWhere is different between the case studies, ranging from 10 km grid resolution for Burgundy to 40 km grid resolution for Europe.

The BeWhere tool can support the development of strategies for establishing biomass delivery chains to reach specific national bioenergy and wider bioeconomy targets. However, before enabling reliable support, it is necessary to feed the tool with data as accurate as possible on many aspects, including biomass cost–supply, existing biomass installations, topography, road and railway network data, heat and power demand, as well as the associated costs/prices. Output of BeWhere can be used as input for further analysis and more precise chain design and evaluation in the LocaGIStics tool. BeWhere will provide to LocaGIStics the locations of the selected production plants, their capacity, and technology chosen.

4.4.2 LocaGIStics

LocaGIStics is a regional assessment tool for biomass delivery chains. This tool can support the user to design optimal biomass supply chains and networks at regional level and analyze in a comparative way (for different biomass supply chains) the spatial implications and the environmental and economic performance. It will take account of the biomass cost–supply, the conversion and pretreatment technology options, and the novel logistical concepts of biomass yards. In relation to environmental impacts, it takes account of indicators and guidelines for assessing the overall sustainability performance for bioeconomy value chains developed.

This tool provides support to regional and local stakeholders in making strategies for best ways to develop their bio-based economy and making use of sustainable local biomass resources potentially available to them. The scale of assessment is to be as detailed as data allows in the case studies for which the tool is developed. The tool was developed and validated in two case study regions.

This tool was developed for the Burgundy case study using input from the Logist'EC project (Fig. 4.3). It enables the users to further design and evaluate in more detail the solutions for additional biomass delivery chains that were proposed by the BeWhere model for Burgundy. These solutions are translated into several biomass delivery designs using the variation in logistical concepts covering transport, pretreatment, and conversion options. The feasibility of every chain design can then be analyzed in a comparative way (for different biomass delivery chains) in relation to environment (greenhouse gas (GHG) emissions and mitigation including land use change emissions, soil C, and an economic minimum acceptable return analysis).

4.4.3 Truck Transport Logistics Model

The Truck Transport Logistics model was compiled in Witness simulation software and combined with an Excel-spreadsheet environment (Fig. 4.4). The Witness simulation platform has previously been utilized to study other logistical concepts of biomass supply (Asikainen, 2001; Karttunen et al., 2012; Windisch et al., 2015). Simulation runs are conducted in Witness,

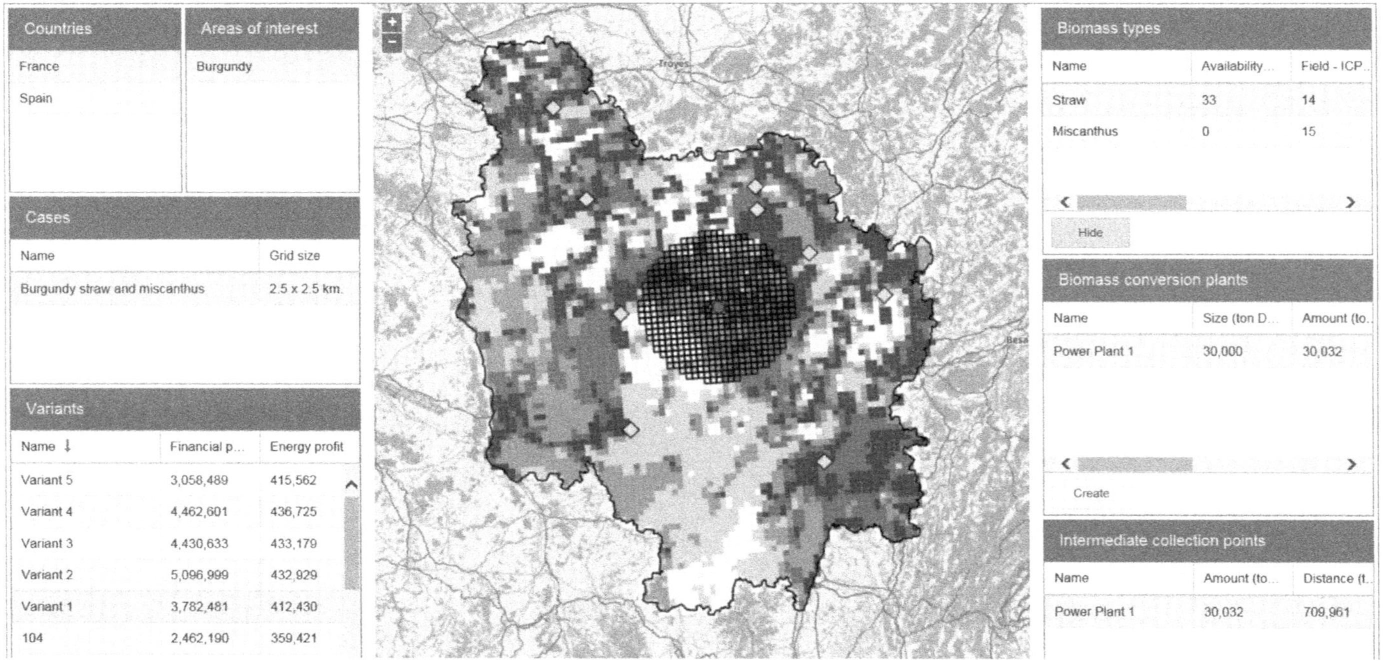

FIGURE 4.3 Example of the interface of LocaGIStics for Burgundy.

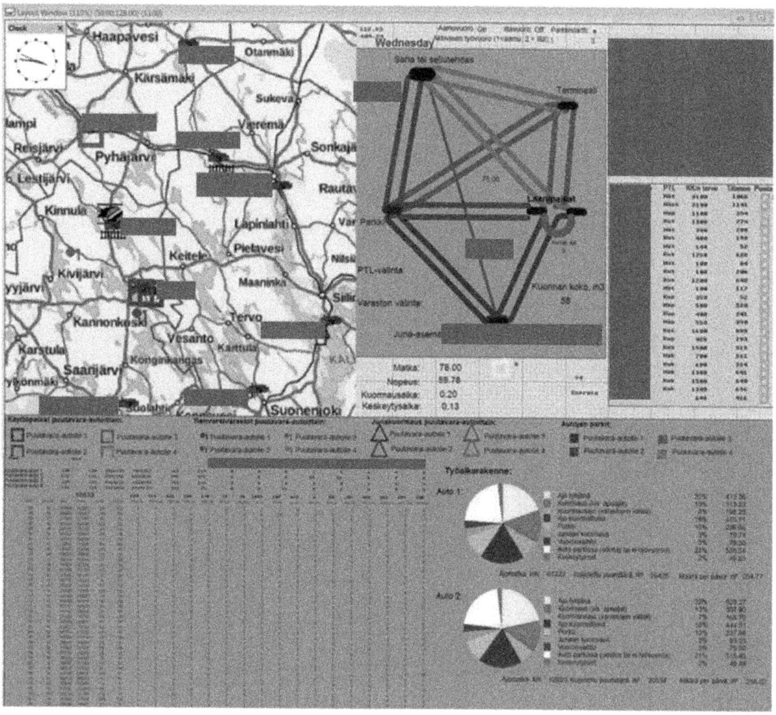

FIGURE 4.4 A screenshot from the Truck Transport Logistics—simulation model in Witness simulation software (the yellow boxes (dark gray boxes in print versions) were used for confidential information).

whereas the Excel-spreadsheet file controls simulation scenario parameters and combines time and performance data from Witness to cost accounting carried out in Excel workbook.

As input data the simulation requires timber storage data at roadside, data of monthly timber demand of mills, parameter input of the work model, and truck characteristics and parameters of timber receptions at the mills. The simulation model reads this information from Excel as initial input data at the beginning of a simulation run. As a result, the model calculates the distribution of working time, transported volumes, utilization rate of capacity, and costs for each logistical concept compared.

4.5 CASE STUDIES

A case-based approach was followed in the S2Biom project, where optimal logistical concepts (conceptual designs) were matched with the specific regional situation (Annevelink et al., 2016b). This was done in three logistical case studies, representative for European conditions in terms of matching geography, technology, and biomass types:

1. Small-scale power production with straw and *Miscanthus* in the Burgundy region in France (Annevelink et al., 2016c)
2. Large-scale power production with straw and with residual woody biomass in the Aragon region in Spain (García-Galindo et al., 2016c)
3. Advanced transport logistics of timber in the Province of Central Finland (Väätäinen et al., 2016).

Data on biomass availability and demand and quality specifications of the conversion technology have been used in combination with data on logistical components and concepts.

4.5.1 Burgundy

4.5.1.1 General Characteristics

The case that was originally described in the Logist'EC project (EU FP7; Perrin et al., 2017) which aimed at developing new or improved technologies for all steps of the logistics chains for biomass supply from energy crops, and to assess their sustainability for small- to large-scale bio-based projects. The supply chain for the small-scale local production of *Miscanthus* pellets for bioenergy consists of *Miscanthus* feedstock cultivation—harvesting as bales or chips—bales stored at the farm—and then transported to the pellet plant—where they are chipped and pelletized—and finally they are transported to a bioenergy power plant.

The focus of the Burgundy case study was put on *Miscanthus* and straw to explore alternative sources of biomass and increase the potential supply overall (Annevelink et al., 2016c). For these types of biomass, the BeWhere model was used to show where there is a possibility to locate a (new) biomass conversion factory specifying the type of technology and size. In this case it was a small-scale combustion power plant to satisfy the energy demand in Burgundy in terms of cost and GHG efficiency based on overall energy (electricity demand) and local biomass availability in different scenarios. LocaGIStics then took the information on the size and type of technology to assess how the organization of the biomass delivery chain should look like in terms of logistical concepts, specifying, for example, alternative user-defined locations for a conversion plant, and for intermediate storage points and pretreatment alternatives given different types and amounts of Burgundy biomass use, etc.

4.5.1.2 Results

Fig. 4.5 presents an output from the BeWhere model with the locations of the bioenergy production plants (black circles). The selected production plants collect the biomass from the closest locations (Fig. 4.5, left side,

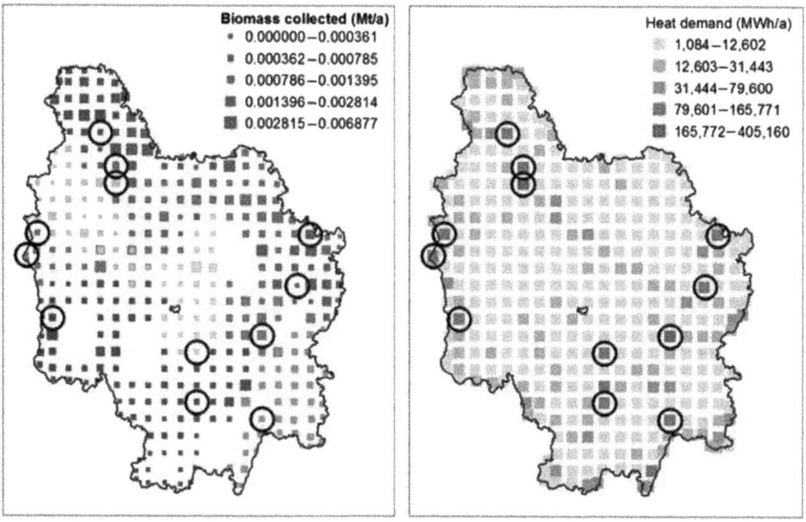

FIGURE 4.5 Example of output of BeWhere for Burgundy.

where a certain color represents the biomass collected for the production plant that is indicated by the circle in that area). The biomass collected is no longer within a circle around the plant, but some optimal distribution around the plant balancing transport cost, availability, and collection cost. The bioenergy production plants produce both heat and power. Heat is assumed to be distributed to a district heating network, and therefore cannot be shipped for distances longer than 30 km.

The right side of Fig. 4.5 presents the location of the plants and at the same time the location of the heat demand in Burgundy. One can notice that the plants are located where the heat demand is the highest. The technology chosen remains the same for all plants as well which is a grate boiler for CHP, with a capacity of 10 MW_{th}. This means that heat demand has a greater impact on the location of the plant than the biomass, which is collected within distances ranging from 70 to 160 km.

The LocaGIStics tool (Fig. 4.3) was used to further detail the biomass value chain of one of these possible locations. It was used as a quality check of the biomass collection, capacity, and therefore the validity of the chosen location. Several logistical concepts have been tested with LocaGIStics in the Burgundy case. These are: (1) mixing different biomass types (straw as a biomass residue and *Miscanthus* as an energy crop), (2) applying pretreatment technology (pelletizing) to densify the material in order to lower the transportation costs and increase handling properties, (3) switching between different types of transport means (truck and walking floor vehicle), and (4) direct delivery to a power plant vs putting an intermediate collection point in the value chain.

4.5.2 Aragón

4.5.2.1 General Characteristics

The case study Aragón has been developed in close cooperation with Forestalia Group, who started the promotion of the Monzón, Zuera, and Erla power plants in Aragón region in 2016 (García-Galindo et al., 2016c). They were scoped to be fed only by means of woody energy crops, but Forestalia Group was also interested in exploring the potential role of other biomass resources. For the presented case, the targeted fuel mix consisted of 70% energy crops and 30% agriculture residues. The aim of the case study consisted of determining the supplying area and at plant gate costs of agricultural residues.

Within this case study, LocaGIStics was used for determining the biomass potential and the supply cost of biomass at plant gate considering the three power plants together and separately. First, the available potential of different agricultural residues was obtained in order to select the main biomass options. The results revealed two main biomass groups of relevance in Aragón region: (1) straw and stalk from annual crops (winter cereals, summer cereals, sunflower) and (2) above- and underground woody biomass from olive, fruit, and vineyard plantations removal. Then, for each biomass option, different supply chains were defined. For herbaceous biomass, the supply chain was direct transport of baled straw and stalks from local providers widespread around the territory (which is called Case 1). For woody residues of plantations removal (which is called Case 2), three different configurations were defined for the supply chain, depending on the size of the field parcels and the concentration. This led to three logistical concepts within Case 2:

Case 2.1: tree stumps dumped by farmers in local collection points
Case 2.2: transport of whole tree uprooted to collection points from intermediate sized fields
Case 2.3: felling of trees, separate handling of above- and underground biomass.

Cases 2.1 and 2.2 involve the costs of a mobile unit with shredder, screening devices, and chippers to the collection points. Case 2.3 involves a direct operation of the mobile unit in the field (fields larger than 2.5 ha).

4.5.2.2 Results

Based on the supply chains, different scenarios were analyzed by LocaGIStics for the two biomass options in terms of the number of power plants and their sites, the biomass availability, the total demand per plant, and the presence of collection points.

Case 1.1 results show that the amount of herbaceous biomass is enough to cover the annual needs of the three power plants in any case. Competition

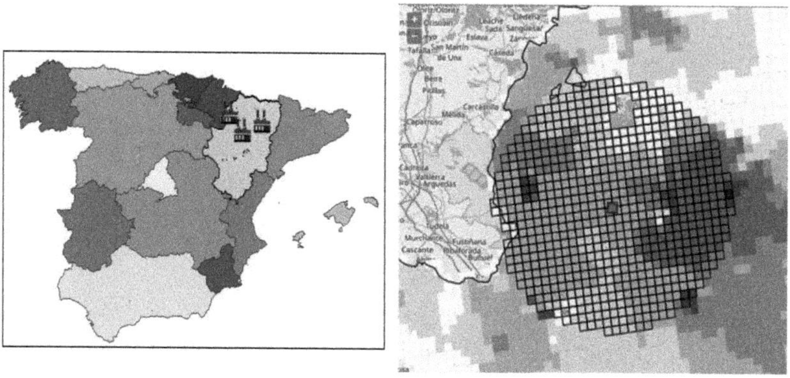

FIGURE 4.6 Example of the sourcing of straw from winter cereals for the Erla power plant in a single plant calculation.

problems appear between the Erla (Fig. 4.6) and Zuera power plants and consequently, biomass collecting distances are longer than those for Monzón. Regarding the final price at gate, the Monzón power plant always showed the minimum value, between 43 and 44 €/t dm. When considering that in a realistic situation not all the biomass in a territory can be contracted, Erla power plant costs increase remarkably when just 50% of the biomass is available.

Regarding Case 2, there is not enough biomass in Aragón to satisfy the demand of a single power plant. Two of the proposed supply chain concepts (Cases 2.1 and 2.2) have a purchase cost higher than the price at gate (considered by Forestalia Group 57 €/t dm). Case 2.3 supply chain was able to mobilize biomass at a price below 57 €/t for the Monzón plant.

In order to complete the analysis, the Zuera power plant (sited in center of Aragón) was studied alone for obtaining the variation of the results regarding the availability percentage from 100% to 25%. The price at gate (€/t) almost did not increase, showing the limited impact of transport in the final costs. However, collected biomass amount was reduced from 60,000 t (100%) to 24,600 t (25%) and the maximum distance increased from 82 to 130 km.

After analyzing the results, it seems clear that in terms of biomass availability and supply chains definitions, the Forestalia Group should focus on straw and stalk as main biomass option. Case 1.1 is technically and economically reliable and there is enough biomass for fulfilling the three power plant fuel requirements.

Regarding wood plantation removal, Case 2.1 and 2.2 supply chains are not profitable. So, the logistical concept should be modified. An option could be to execute the minimum operations with mobile units, e.g., by using a primary crusher for comminution operations at the field side, and then the shredded material could be transported directly to the power plant, where the operational costs with static screening and chipping machines will be lower in comparison to mobile units.

4.5.3 Finland

4.5.3.1 General Characteristics

In the Finnish case study, saw logs, pulp wood, and de-limbed energy wood stems were transported to end-use facilities (Väätäinen et al., 2016). In total, 25 different timber assortments were included in the supply chain of this case. Currently, each timber assortment is transported as single-assortment loads to the end-use facility. Due to the small volume of individual assortments in a roadside storage, the timber trucks often have to collect timber from several roadside storages to obtain a full load. This kind of driving between piles at different roadside storages and setup times at these piles are relatively time-consuming elements in the whole transport cycle. Therefore, a scenario with a multiassortment load option was introduced to the case study. The multiassortment load opportunity is only available for timber assortments, which are all transported to a same end-use facility.

In the Finnish case study, the Truck Transport Logistics model included four trucks operating in Central Finland and supplying timber to 12 end-use facilities being eight saw mills, two pulp mills, and two train loading terminals. A simulation run covered a period of 1 year. Each scenario was simulated by five stochastic repetitions and the average values of these five repetitions were used for calculating the result data of a certain scenario. Two simulation scenario sets were simulated in Finnish case study. The business as usual scenario corresponded to timber transports with the single-assortment load method, whereas the multiassortment scenario included the multiassortment load transports.

4.5.3.2 Results

The multiassortment method decreased the time consumption particularly for driving between piles. The number of rides between piles was 1152 times in the single-assortment load method, whereas in the multiassortment load method driving between piles was 794 times. This is a 31% decrease. On the other hand, the off-shift time increased in the multiassortment load method mainly because of roadside storage was finished earlier during June before the holiday month (July) in the multiassortment load method.

The multiassortment load method was on average 3.3% cheaper than the single-assortment load method. In addition, the driving performance—presented as solid-m^3 of timber per 100 km—was 4.0% higher with the multiassortment load method.

The Truck Transport Logistics model has proved to express well the behavior of truck transports of timber. The model could be used for supporting decisions on enhancing the transport operations. Compared with the other logistical assessment methods in S2Biom, i.e., BeWhere and LocaGIStics, Truck Transport Logistics is the most detailed one. This means that it can simulate the operation of real logistical chains and even the interactions between logistical components and stochasticity can be taken into account. The downside of this ability is that very detailed input data, which is not always available, is needed. The model also needs to be tailored to the operating environment which requires expertise.

The multiassortment load method that was studied in the Finnish case study offers a nearly 4% improvement potential for the transport economy compared to the prevailing single-assortment load method. Small assortment piles at roadsides cause difficulties in efficient timber transport due to driving between piles and the need of loading many small piles for filling the entire load space. The multiassortment load method decreases drastically the number of rides between piles and, therefore, improves performance of the fleet.

REFERENCES

Alakangas, E., Virkkunen, M., 2007. Biomass fuel supply chains for solid biofuels. EUBIONET2 project (EIE/04/065/S07.38628), 32 pp.

Alakangas, E., Routa, J., Asikainen, A., Nordfjell, T. (Eds.), 2015. Innovative, effective and sustainable technology and logistics for forest residual biomass—summary of the INFRES project results, 40 pp.

Annevelink, E., de Groot, H., Shah, N., Giarola, S., Pantaleo, M., Anttila, P., et al., 2014. Review of the main logistical components. S2Biom project, Deliverable D3.1, 65 pp.

Annevelink, E., de Groot, H., Shah, N., Giarola, S., Pantaleo, M., Anttila, P., et al., 2015a. S2Biom database with logistical components of the biomass value chain. Proceedings of the 23d European Biomass Conference and Exhibition, June 1–5, 2015, Vienna, Austria, pp. 339–343.

Annevelink, E., Staritsky, I., Krajnc, N., Lammens, T., Sanchez Gonzalez, D., Lenz, K., et al., 2015b. Logistical concepts. S2Biom project, Deliverable D3.2, 83 pp.

Annevelink, E., Elbersen, B., Leduc, S., Staritsky, I., 2016a. Formalized stepwise approach for implementing logistical concepts using BeWhere and LocaGIStics. S2Biom project, Deliverable D3.5, 41 pp.

Annevelink, E., Gabrielle, B., Carozzi, M., García Galindo, D., Espatolero, S., Izquierdo, M., et al., 2016b. Results logistical case studies, Deliverable D3.4 + D3.6 Cover report, 23 pp.

Annevelink, E., Gabrielle, B., Carozzi, M., Staritsky, I., Vanmeulebrouk, B., Elbersen, B., et al., 2016c. Results logistical case study Burgundy. S2Biom project, Deliverable D3.4 + D3.6 Annex 1, 48 pp.

Annevelink, E., Staritsky, I., Vanmeulebrouk, B., Elbersen, B., Gabrielle, B., 2016d. LocaGIStics: design & evaluation of regional biomass delivery chains. Proceedings of the 24th European Biomass Conference and Exhibition, June 6–9, 2016, Amsterdam, pp. 174–175.

Annevelink, E., Staritsky, I., Krajnc, N., Lammens, T., Sanchez Gonzalez, D., Lenz, K., et al., 2016e. S2Biom survey of logistical concepts. Proceedings of the 24th European Biomass Conference and Exhibition, June 6–9, 2016, Amsterdam, pp. 108–113.

Asikainen, A., 2001. Simulation of logging and barge transport of wood from forests on islands. Int. J. Forest Eng. 2 (2).

Gabrielle, B., Flatberg, T., Perrin, A., Wohlfahrt, J., Bjørkwoll, T., Echevarria-Goni, I., et al., 2015. Improving logistics for biomass supply from energy crops in Europe: main results from the Logist'EC project. Proceedings of the 23rd EU Biomass Conference and Exhibit, June 3, 2105, Vienna, ETA, pp. 1–5.

García-Galindo, D., López, E., Gómez-Palmero, M., Sebastián, F., Gebresenbet, G., Jirjis, R., et al., 2016a. EUROPRUNING project: summary of final result. Oral Presentation at the 24th European Biomass Conference, Amsterdam, June 2016.

García-Galindo, D., Gómez-Palmero, M., López, E., Sebastián, F., Jirjis, R., Gebresenbet, G., et al., 2016b. Agricultural pruning harvesting demonstrations in Germany, France and Spain. Lessons learned and recommendations. Poster at the 24th European Biomass Conference, Amsterdam, June 2016.

García-Galindo, D., Espatolero, S., Izquierdo, M., Staritsky, I., Vanmeulebrouk B., Annevelink, E., 2016c. Results logistical case studies Aragon. S2Biom project, Deliverable D3.4 + D3.6 Annex 2, 77 pp.

Karttunen, K., Väätäinen, K., Asikainen, A., Ranta, T., 2012. The operational efficiency of waterway transport of forest chips on Finland's Lake Saimaa. Silva Fennica 46 (3), 395–413.

Kisslinger, T., 2004. Biomass logistics at Essent Energie's co-firing power plant. Port Technol. Int.2–4.

Lanner, 2016. <http://www.lanner.com/technology/witness-simulation-software.html>.

Leduc, S., 2009. Development of an optimization model for the location of biofuel production plants. Doctoral thesis, Luleå University of Technology, Sweden, 71 pp. + 6 Annexes.

Leduc, S., Wetterlund, E., Dotzauer, E., Kindermann, G., 2012. CHP or biofuel production in Europe? Energ. Procedia 20, 40–49.

Ljungblom, L., 2011. Remote forest fuel for huge city CHP. Bioenerg. Int. 51 (3), 10–11.

Natarajan, K., Leduc, S., Pelkonen, P., Tomppo, E., Dotzauer, E., 2011. Optimal locations for methanol and CHP production in Eastern Finland. Bioenerg. Res. 5 (2), 412–423.

Nivala, M., Anttila, P., Laitila, J., 2015. A GIS-based comparison of long-distance supply of energy wood for future needs from young forests to the coast of Finland. Int. J. Forest Eng. 26 (3), 185–202.

Perrin, A., Wohlfahrt, J., Flatberg, T., Morandi, F., De la Rua, C., Østergård, H., et al., 2017. Integrated design and assessment of innovative biomass supply chain logistics: a case-study on miscanthus in Burgundy, France. In revision for Applied Energy.

Routa, J., Asikainen, A., Björheden, R., Laitila, J., Röser, D., 2012. Forest energy procurement—state of the art in Finland and Sweden. Wiley Interdisc. Rev. WIREs Energ. Environ. 2012. Available from: http://dx.doi.org/10.1002/wene.24.

Sanders, J.P.M., Annevelink, E., van der Hoeven, D., 2009. The development of biocommodities and the role of North West European ports in biomass chains. Biofpr (May/June), 395–409.

Strandström, M., 2015. Timber harvesting and long-distance transportation of roundwood 2014. Metsätehon tuloskalvosarja 7b/2015. Available from: <http://www.metsateho.fi/timber_harvesting_and_long_distance_transportation_of_roundwood_2014>.

Stodafor, 2004.Technical guide on harvesting and conservation of storm damaged timber, CTBA, 103 pp. <www.unece.org/fileadmin/dam/timber/storm/Stodafor%20manual/stodafor-manual-2004.pdf>.

Tahvanainen, T., Anttila, P., 2011. Supply chain cost analysis of long-distance transportation of energy wood in Finland. Biomass Bioenergy 35 (8), 3360–3375.

Väätäinen, K., Anttila, P., Annevelink, E., 2016. Logistical case study in Finland. S2Biom project, Deliverable D3.4 + D3.6 Annex 3, 23 pp.

Wetterlund, E., Leduc, S., Dotzauer, E., Kindermann, G., 2013. Optimal use of forest residues in Europe under different policies—second generation biofuels versus combined heat and power. Biomass Conversion Biorefin. 3 (1), 3–16.

Windisch, J., Väätäinen, K., Anttila, P., Nivala, M., Laitila, J., Asikainen, A., et al., 2015. Discrete-event simulation of an information-based raw material allocation process for increasing the efficiency of an energy wood supply chain. Appl. Energ. 149, 315–325.

Woodenergy, 2014. Wood as a fuel. <www.woodenergy.ie/woodasafuel>.

Woodfuel Handbook, 2009. <http://nuke.biomasstradecentres.eu/Deliverables/D21WoodFuelsHandbook/tabid/58/Default.aspx>

Chapter 5

Modeling Demand for Bio-Based Sectors

Marc Londo, Joost van Stralen, Carolien Kraan, Hamid Mozaffarian and Ayla Uslu

5.1 INTRODUCTION

In order to understand the potential role of biomass in a future sustainable energy and material economy, it is important to understand the key complexities of biomass applications. This is a complex matter, as biomass has a wide variety of resources, conversion routes, and application markets, not only in energy but also in chemicals and materials. Standard integrated assessment models (IAMs) (Loulou et al., 2004, 2016) are well able to address questions on the role of biomass at a more systemic level (e.g., the competition of bio-based routes vs other options for low-carbon power, heat, and clean transport). However, given their inherently generic description of biomass chains, they face difficulties in addressing questions that require a more detailed coverage of bio-based supply chains and markets; think of more geographically detailed analyses, deeper insights in trade flows of biomass and bio-based products between countries, and specific technology developments or innovations in feedstock mobilization.

Therefore, there is a need for dedicated modeling of bio-based sectors, covering the full range from feedstocks to products, with more details regarding chains and technologies, and geographically more detailed. This chapter reviews key developments on this respect, taking the developments and step-by-step improvements of the BIOTRANS/RESolve-Biomass model of ECN as a case. This is an optimization model providing least-cost mixes of biomass chains meeting a given demand-defined objective, given various possible constraints along the chain. The model was developed and further elaborated in a chain of EU projects. Each time, the specific objectives of the assessments were different:

- In the VIEWLS project (2003–05, funded by the EU Intelligent Energy Europe (IEE) programme), the BioTrans model was first developed. Key

Modeling and Optimization of Biomass Supply Chains.
DOI: http://dx.doi.org/10.1016/B978-0-12-812303-4.00005-7

objective was to explore the least-cost biofuel mix meeting given (energy-defined) targets for 2020 and 2030 (Wakker et al., 2005). The results fed into the general discussion by then on the impacts of the Biofuels Directive and the first preparations of the Renewable Energy Directive (RED).

- In the REFUEL project (2006–08, IEE), BioTrans was further refined, also in terms of input data. Again, it provided least-cost biofuel mixes given energy-defined targets and various other possible constraints for 2020 and 2030 (Londo et al., 2008; Londo, 2008). The results fed into further preparations of the RED.
- In the Elobio project (2007–10, IEE), specific analyses with BioTrans were done in order to explore options for reducing the investment risk barrier for advanced biofuels (Bole et al., 2010). The results fed into the discussions on food vs feed competition and biofuels possibly disturbing food markets (which was a general topic in the project), and the perspectives of advanced biofuels for reducing these disturbances.
- In the Biomass Futures project (2009–11, IEE), the focus was on checking the feasibility of the biomass ambitions of member states as formulated in the National Renewable Energy Action Plans (NREAPs), and provide suggestions for improvement (Uslu et al., 2012). The results were used in the further refining of the NREAPs over the years.
- In the Biomass Policies project (2013–16, IEE), RESolve-Biomass was applied specifically to assess the impact of various policy options to further mobilize biomass and ramp up bioenergy development at national level, exporting best practices from one member state to the other (van Stralen et al., 2016; Uslu and van Stralen, 2016). The results were disseminated by various energy agencies participating in the project, providing concrete advice to their governments.
- In the S2Biom project (2013–16, FP7), RESolve-Biomass was used to assess the least-cost mix and feedstock implications of meeting national bioenergy objectives and ambitions for bio-based chemicals (van Stralen et al., 2016). The results were used to provide more insight in the synergies and competition effects between bio-based energy and chemicals.

This chapter reviews the progress made in the integrated modeling of biomass demand and supply systems over the course of these projects, and the specific changes and improvements that were made along the way. The structure is as follows:

- In Section 5.2, we go into the progress in assessments of markets for bio-based energy carriers and products; essential input to the models
- In Section 5.3, we describe progress in the modeling approach itself: its modeling mechanics, refinements in the modeling approach, including level of detail, coverage of products and energy carriers, and other characteristics

- In Section 5.4, we discuss developments in data quality and scenarios
- The chapter ends with conclusions and recommendations for future improvements.

5.2 MARKETS FOR BIO-BASED ENERGY AND PRODUCTS: CHARACTERISTICS AND OUTLOOK

The RESolve-Biomass model is an optimization model that requires the demand for final bioenergy and bioproducts as one of its key (exogenous) inputs. The evolution of the methods for assessing demand sectors is summarized in Section 5.1. For the work in S2Biom the demand for different bioenergy carriers—electricity, heat, and liquid fuels—was updated based on a market analyses of the bioelectricity, bioheat, and biofuel markets in all countries included in the geographic scope. These will be elaborated upon in Section 5.2.2. The RESolve-Biomass model was also extended to include the demand for the most promising bioproducts for purposes in the chemical sector, of which the corresponding market analysis is described in Section 5.2.3.

5.2.1 Demand Sectors and Their Assessment Methods

Table 5.1 provides in retrospect the development of the inclusion of demand sectors in the modeling for the various projects in which BIOTRANS/RESolve-Biomass was used. It illustrates the widening of its scope, starting from biofuels only to a much wider ranges of bio-based applications. This also means that competition and synergy effects are gradually better modeled. However, the benefits of complex biorefinery systems that provide a wide array of outputs are not yet within the modeling scope.

In terms of reference data, all projects relied on the most recent policy ambitions of their time, and studies that converted those ambitions into demand for the various bio-based energy carriers. For bio-based chemicals, the demand assessment is still more indicative as there are no specific quantitative policy targets. For the reference scenario of the overall energy system, the Commission's publications on the basis of PRIMES analyses have always been the basis.

5.2.2 Energy Demand Assessment in S2Biom

In the most recent application of RESolve-Biomass, the final energy demand to be covered by bio-based options assessment was taken from country-specific data sources. While the model uses these country-specific data, we present them both in total (Fig. 5.1) and for the six specific regions defined in the project (Fig. 5.2). As mentioned, the key source for these data is the GREEN-X modeling work provided in the BETTER project (Liebmann et al., 2015), analyzing the least-cost pathway to reach the 20% and 27%

TABLE 5.1 Assessment of Demand for Bio-Based Energy Carriers and Products: Coverage and Assumptions in the Course of the Projects

Project	Demand Sectors Covered[a,b]	Combinations Modeled	Sources for Demand Assessment	Overall Energy Reference Scenario
S2Biom (2016) (Kraan, 2015; Mozaffarian et al., 2016)	HEFC	CHP, heat, and power as coproducts of some chains for biofuels and bio-based chemicals	HEF: GREEN-X modeling, and NREAPS, consistent with RED targets (2020) and climate package (2030) C: Market review (2020/30)	2030 Energy and Climate package impact assessment, GHG40 scenario
Biomass Policies (2016) (van Stralen et al., 2016; Uslu and van Stralen, 2016)	HEF	CHP, heat, and power as coproducts of some biofuels chains	HEF: GREEN-X modeling and NREAPS, consistent with RED targets (2020) and climate package (2030). However, for HE further refined by interaction with the RESolve-E model	2030 Energy and Climate package impact assessment, GHG40 scenario
Biomass Futures (2012) (van Stralen et al., 2012)	HEF	CHP, heat, and power as coproducts of some biofuels chains	HEF: NREAPS and PRIMES 2009 Ref. For E further refined by interaction with the RESolve-E and -H/C models	PRIMES 2009 reference
Elobio (2010) (Bole et al., 2010)	F	Heat and power as coproducts of some biofuels chains	2020: RED (10%) 2030: project-defined ambition (15%)	PRIMES 2006 reference

(*Continued*)

TABLE 5.1 (Continued)

Project	Demand Sectors Covered[a,b]	Combinations Modeled	Sources for Demand Assessment	Overall Energy Reference Scenario
REFUEL (2008) (Londo et al., 2008)	F	Heat and power as coproducts of some biofuels chains	2020: RED[c] (10%) 2030: project-defined ambition (25%)	PRIMES 2006 reference
Viewls (2005) (van Thuijl et al., 2005)	F	None	2010: BFD[c] (5.75%) 2030: project-defined ambition (20%)	PRIMES 2003 reference

[a]*Demand sectors: H, bioheat; E, bioelectricity; F, biofuels; C, bio-based chemicals.*
[b]*The latest version of RESolve-Biomass also covers biofuels in aviation and maritime (de Jong et al., 2017).*
[c]*RED: Renewable Energy Directive; BFD: Biofuels Directive.*

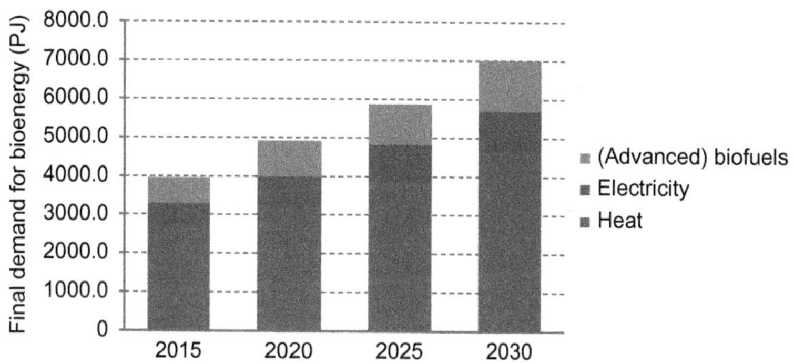

FIGURE 5.1 Total final demand for bio-based energy carriers in the S2Biom scenarios (van Stralen et al., 2016).

RES ambitions of the EU in 2020 and 2030, respectively, in which the 20% target for 2020 includes specific national targets. This analysis assumed that Turkey would not become an EU member nor Energy Community (EnC) affiliate (thus having a weaker RES target), and with cooperation within the EU possible, but no cooperation between the EU28 and non-EU28 countries (specific name of the scenario: Default-TRLowAmbition (weak)).

For Moldova and Ukraine, who were not part of the GREEN-X modeling, use was made of data from their respective NREAPs. This was further extrapolated to 2030 using the average growth in EU13 countries (Bulgaria, Cyprus, Czech Republic, Estonia, Croatia, Hungary, Lithuania, Latvia, Malta, Poland, Romania, Slovenia, and Slovakia).

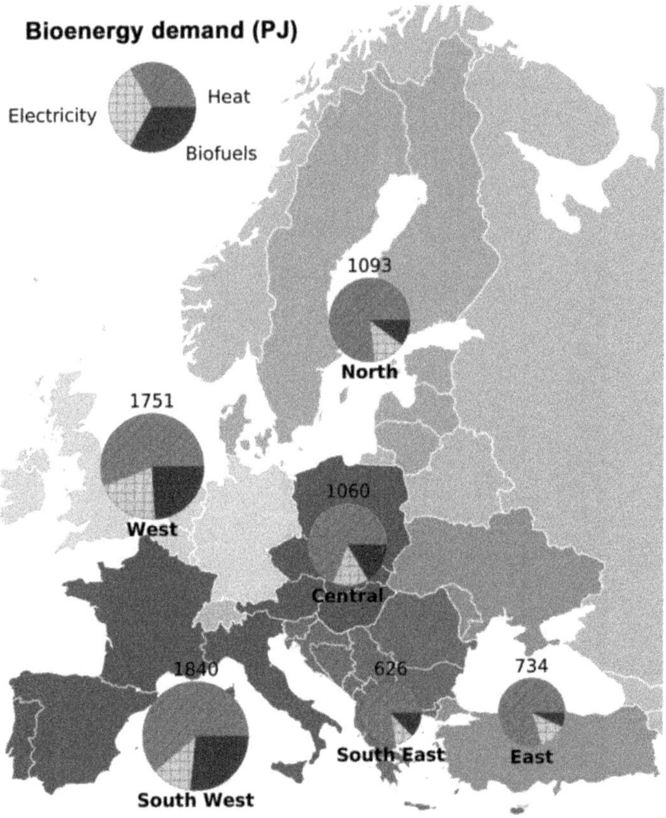

FIGURE 5.2 Final demand for bioenergy in the EU28 + region (in PJ) in the S2Biom assessment (van Stralen et al., 2016).

The data clearly show an ongoing increase in final demand for bio-based energy carriers, with bio-based heat remaining the largest demand sector, but relative growth being stronger in bio-based electricity and biofuels.

The regional specification of final bioenergy demand (Fig. 5.2) indicates that the heat dominance is strongest in regions in which decentralized biomass is widely available (such as the North, Central, and two Eastern regions) and decentralize use in rural areas is widespread. Quite strikingly, the relative role of bio-based electricity is largest in the West region. Biofuels have their largest share in the South West region.

5.2.3 Bio-Based Chemicals

The inclusion of future demand for bio-based chemicals was first introduced in S2Biom. For this, an extensive market review was executed:

- First, a preselection of product–market combinations (PMCs) was done in order to identify PMCs that could generate a relatively significant demand of biomass. This implies a focus on relatively bulky, base-grade bio-based chemicals and platform molecules, and puts less attention to high-value, low-volume bio-based chemicals.

- From this preselection, only PMCs were selected that (can) use lignocellulosic biomass as a feedstock, as the S2Biom project focused on this type of biomass. (In the full market review, demand for chemicals based on other feedstocks (such as vegetable oils and starches) was still collected as by-catch.) This selection, including the PMCs related to bio-based energy, is given in Table 5.2.
- From this selection of lignocellulose-based PMCs, a market review was performed for each PMC.

Full details of the market review for bio-based chemicals can be found in Mozaffarian et al. (2016). Some specific remarks are as follows:

- Novel chemical routes taking C5 and C6 sugars show wide diversity of molecules to be produced from them. The literature source material allowed us to make a generic estimate of total demand of C6 and C5 sugars, broken down into demand for polymers and plastics, surfactants, and solvents. This provided a basis for an indication of corresponding biomass demand, without the need to go into full details of all underlying production pathways (Pantaleo, 2015).
- Biomethane, biomethanol, and hydrogen are special cases as they can be used both as an energy carrier (e.g., for heating or transport) and as a

TABLE 5.2 Complete Set of Bio-Based PMCs (for Energy and Chemicals) for Market Review in S2Biom (Mozaffarian et al., 2016)

	Product	Market
1	Heat	District heating
2	Electricity	Power market
3	Advanced biofuels	Transport fuel
4	C6 sugars	C6 chemistry: polymers and plastics, others
5	C5 sugars	C5 chemistry: polymers and plastics, others
6	Biomethane	Grid, transport
7	BTX[a]	Petrochemical industry
8	Methanol	Transport, chemical industry
9	Hydrogen	Transport, (petro)chemical industry
10	Ethylene	(Petro)chemical industry

[a]*Benzene, toluene, xylene.*

feedstock for chemicals. In this review, we reviewed and combined demand projections for both functions.

- BTX (benzene, toluene, and xylene) is a group of chemicals that can be produced in low-temperature gasification processes, together with methane. As these are always produced jointly, we identified them as a group.

The indications of final demand for these bio-based chemicals by 2020 and 2030 were first converted into corresponding primary demand for biomass, taking into account conversion efficiencies and losses along the chain. This resulting biomass demand for the various bio-based chemicals applications is shown in Fig. 5.3. Later, we integrated the conversion technologies into bio-based chemicals into the model, also introducing the related conversion efficiencies in it. For biomethane, demand is expected to be dominated by its function as an energy carrier, and its lignocellulosic biomass demand will be relatively large in comparison with the chemicals PMCs: up to 80 PJ in 2020 and 280 PJ in 2030.

The market assessment indicates that the PMCs that stem from thermochemical pathways (ethylene, hydrogen, methanol, BTX) may create more biomass demand than the three biochemical routes through C5/6 sugars from lignocellulose (surfactants, solvent, and polymers and plastics). But the significant ranges between low and high scenarios also indicate that there is considerable uncertainty in these estimates, mainly because technologies are still under development, and specific support policies for bio-based chemicals are not as widespread as support policies for bio-based energy.

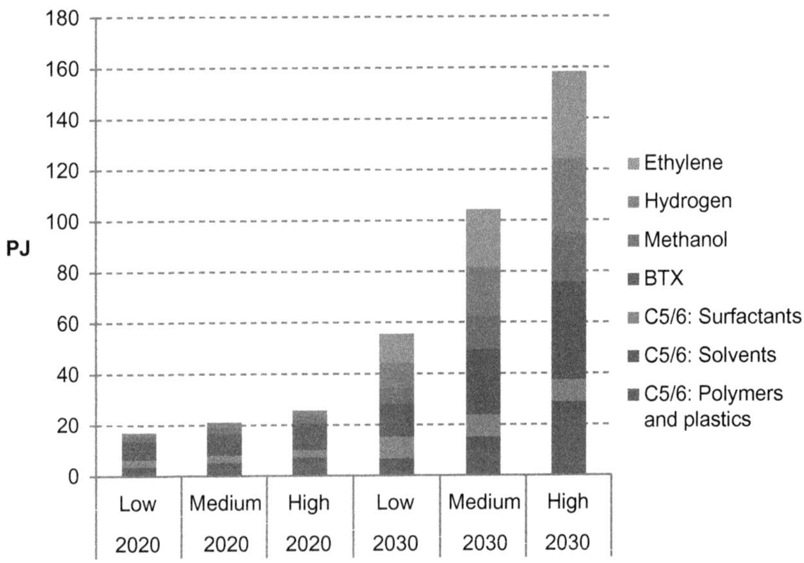

FIGURE 5.3 Projected lignocellulosic biomass demand for the selected bio-based chemicals PMCs, excluding biomethane (Mozaffarian et al., 2016).

5.3 MODELING FRAMEWORK

The RESolve-Biomass determines the least-cost configuration of the entire bio-based production chain, given demand projections for biofuels, bioelectricity, bioheat, and biochemicals (more specifically, chemicals made from lignocellulosic biomass), and elaborate details on biomass potentials and conversion technologies (including technological progress), under several possible conditions and constraints (van Stralen et al., 2012; van Tilburg et al., 2005; Lensink et al., 2007). By doing so it mimics the competition among these four sectors for the same resources. The model optimizes from year to year: in contrast to some other models that assess the optimal pathway to an end year.

The RESolve-Biomass model includes raw feedstock production, (pre) processing, conversion, transport, and distribution. One of the most important features of the RESolve-Biomass model is the ability to link the national production chains, allowing for international trade in biomass, intermediates, and final bio-based products. By allowing trade, the future cost of bioenergy and biochemicals can be approached in a much more realistic way than when each country is evaluated separately. Its most recent functional description stems from the Biomass Futures project (van Stralen et al., 2012), most recent updates and expansions can be found in the S2Biom Integrated Assessment (van Stralen et al., 2016).

5.3.1 Key Modeling Features Developed Over Time

An overview of key innovations along the development process of the model is given in Table 5.3. It illustrates that an existing model or tool is never directly fit to address new questions: every time, key innovations or adaptations were needed in order to adequately address new questions. On the other hand, the basic architecture of the model appeared to be sufficiently flexible to absorb these adaptations and remain a useful tool. This illustrates the need for pragmatic handling of this kind of models and tools: on the one hand, they can never be used right away for a variety of new questions, but when handled with care, the core of their approach can be versatile.

5.3.2 Key Characteristics Throughout the Model's Lifetime

While the model was developed and adapted over time, some characteristics remained identical or were a permanent point of attention. Essentials are:

1. A description of the full chain of bio-based energy and materials, from feedstock to end use, in a step-by-step approach. This allows the model to choose optimal combinations of feedstocks, logistics (including international trade), conversion, logistics again (with trade options), and end use.

TABLE 5.3 Modeling Objectives and Key Innovations to BioTrans/RESolve-Biomass in the Course of Time

Project (Final Year) Model Name	Demand Sectors Covered[a]	Prime Modeling Objective	Key Innovations in Comparison to Earlier Versions
S2Biom (2016) RESolve-Biomass (van Stralen et al., 2016)[b]	HEFC	Least-cost mix of chains meeting given demand for bio-based energy and chemicals	• Ability to show impacts on marginal costs of biomass resources used • Refinement of technology descriptions and of logistics costs calculations (including oil price dependency and bulk density)
Biomass Policies (2016) RESolve-Biomass (van Stralen et al., 2016)	HEF	Assessment of impacts of specific national biomass policies	• Ability to analyze policy impacts on a more national level • Inclusion of sunk costs and growth restrictions to avoid unrealistically fast changes in feedstocks and conversion technologies
Biomass Futures (2012) RESolve-Biomass (van Stralen et al., 2012)	HEF	Feasibility of NREAP bioenergy ambitions	• Inclusion of heat and electricity sectors • Ability to optimize bioheat and biopower in competition with other renewable power and heat options (RESolve-E, RESolve-H) • Inclusion of feed-in tariffs as possible support policies • Refinement of GHG calculations • Inclusion of statistical data for a better reproduction of historical trends
Elobio (2010) BioTrans (Bole et al., 2010)	F	Assessment of specific policies on advanced biofuels	• Refinement of the impact of technology risk on capital cost • Inclusion of double counting

(Continued)

TABLE 5.3 (Continued)

Project (Final Year) Model Name	Demand Sectors Covered[a]	Prime Modeling Objective	Key Innovations in Comparison to Earlier Versions
REFUEL (2008) BioTrans (Lensink et al., 2007)	F	Least-cost biofuels mix under various policy constraints	• Explicit modeling of technology learning • Inclusion of various policy constraints and CO_2 pricing • Streamlining of overall model approach, reduction of calculation time
Viewls (2005) BioTrans (van Tilburg et al., 2005)	F	Least-cost biofuels mix	None, version 1.0

[a]*Demand sectors: H, Bioheat; E, bioelectricity; F, biofuels; C, bio-based chemicals.*
[b]*In the same period, biofuels in aviation and shipping were also added to the model framework (de Jong et al., 2017).*

2. Statistical data have been updated. A meaningful analysis of future trends starts with a good reproduction of historical trends. Statistical data have been used to calibrate the model.
3. The prevention of radical changes from year to year ("flip-flop behavior"). This is a feature frequently observed in models that optimize from year to year. By introducing logical constraints throughout the generations of the model (the impact of sunk costs, maximum rates of feedstock and technology growth, and a vintage approach), the model has gradually become more stable in this respect.
4. The impact of technology learning has always been an important feature of the model (de Wit et al., 2010). For established technologies, the model applies (endogenous) cost reductions through a learning curve approach. For novel technologies, for which such an approach is less adequate, technology cost reductions are dependent on a maximum rate of upscaling and a related scale factor for the impact on costs.
5. As most bio-based options will need to be incentives with policies, the model has always had ample attention for the impacts of policies. The basis mechanism (least-cost solution for a given volume target) most closely resembles obligation policies, but the model can also assess the impact of feed-in support, and accompanying measures such as investment subsidies, specific support for feedstock mobilization, caps on specific biofuels, and the double counting mechanism for advanced biofuels.

5.3.3 Key Model Strengths

Throughout our work with the model, we experienced several key strengths of it:

1. It covers a relatively fine level of detail, in terms of feedstocks, conversion technologies, and the logistics in-between (see Fig. 5.4 for an example of technology mixes). Especially compared with other integrated energy models or models for all renewable energy options for the EU level, RESolve-Biomass offers more detailed insights in the bio-based sectors.
2. The fact that bio-based chemicals are included in its most recent version is an innovation we have not seen yet in other models in the field; the same holds for the inclusion of end-use and related supply chain elements for biofuels in aviation and marine.
3. Import and export between countries can be explicitly shown, for feedstocks, intermediates, and final products (see Fig. 5.5 for an example of S2Biom). Import/export between countries is explicitly modeled.

5.3.4 Key Model Limitations

As with any model, knowledge of its limitations is pivotal for adequately using it and interpreting its outcomes. Key limitations we see in the (current version) of RESolve-Biomass are:

1. The cost–supply data of the various feedstocks are fixed, although they can change of the years. Especially for feedstocks that can also be used for other applications than energy (think of food crops), this is a strong simplification: demand-side dynamics in other markets than energy and chemicals will directly influence feedstock prices. However, through the information on marginal feedstock costs, it can now checked to what extent these costs deviate from current or expected market prices.
2. The model does not include explicit land use dynamics or competition with other types of land use. This also means that it is not suited to assess complex systemic effects such as indirect land use change (ILUC), for this macroeconomic models land use change models need to be used in combination.
3. The demand for final bio-based energy and chemicals is fixed, which means that the model does not have any dynamic interaction:
   - With other RES options such as wind, solar, and geothermal. This interaction, however, can be done in a simplified way by transferring marginal costs of bio-based options from RESolve-Biomass to RESolve-E and RESolve-H, in which they compete with other renewable options for electricity and heat, respectively. However, these models are less elaborated and up-to-date in data terms.

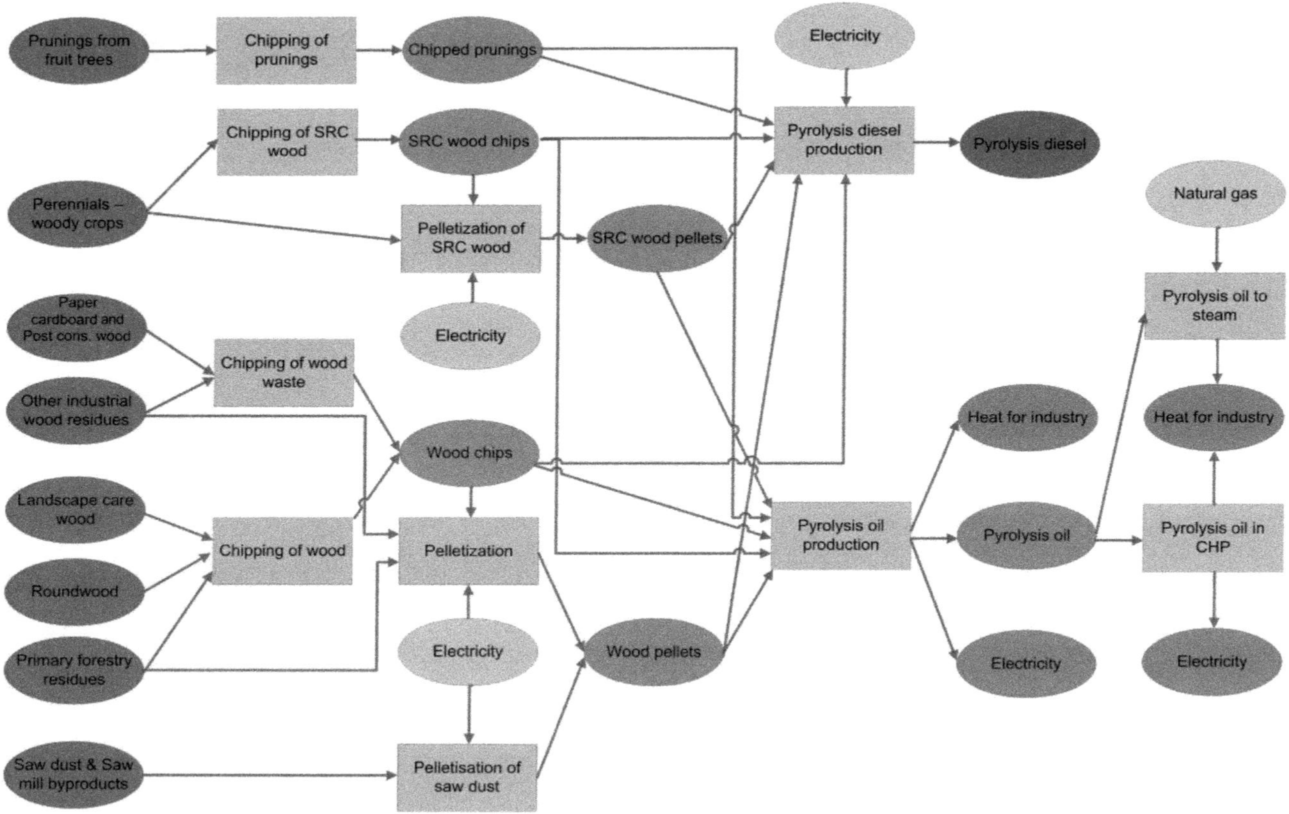

FIGURE 5.4 The set of technologies and feedstocks related to pyrolysis in RESolve-Biomass (van Stralen et al., 2016).

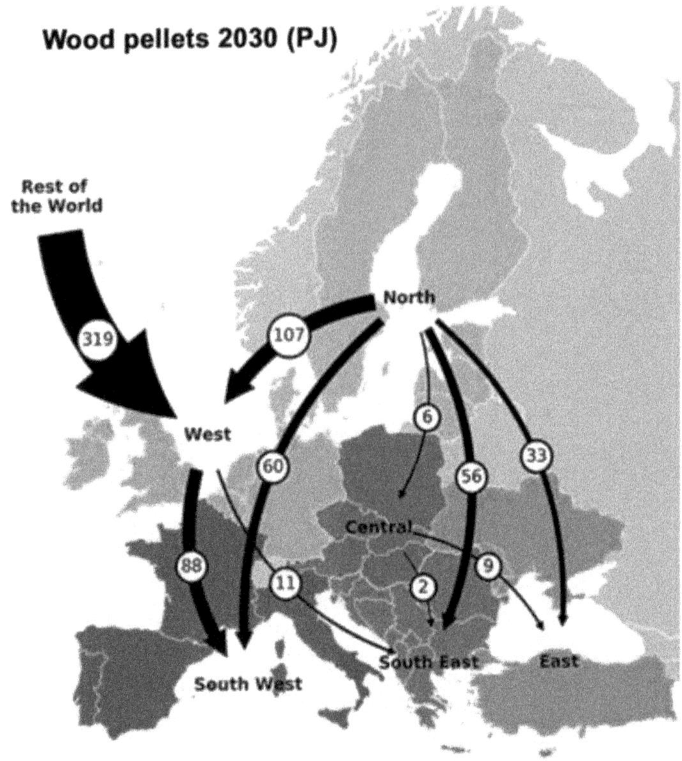

FIGURE 5.5 Trade in wood pellets in the EU and from Rest of the World in the S2Biom HC scenario (van Stralen et al., 2016).

- With fossil options. Although fossil energy prices are included in the model (also to calculate the overall system costs compared to a reference with more fossil energy and chemicals use), they are not affected by a decreasing use of them because of renewables growth. With low shares of renewables this is well acceptable, but with the foreseen more significant role of renewables this becomes problematic.
- With other GHG reduction options like Electric vehicles (EVs), CO_2 Capture and Storage (CCS), and energy savings. The key denominator of the model is energy (euros per PJ of final energy). This means that the model is not fit to find optimal balances in terms of least costs per ton avoided CO_2, although the model contains GHG data for most of its possible chains.

5.4 DATA AND SCENARIO ASSUMPTIONS

Any modeling exercise depends on the quality of the data used, and its relevance also strongly relates to the actual questions addressed and scenarios considered. In this section we discuss the key data used in the model versions (Section 5.4.1) and the scenarios addressed (Section 5.4.2).

5.4.1 Input Data to the RESolve Modeling

The key input data to conduct scenario analysis through the RESolve model are:

- the biomass cost–supply data;
- the techno-economic data related to conversion technologies (both pretreatment and final conversion);
- input data for logistics of feedstocks, intermediate products and fuels;
- end-use costs.

In this section, we focus on the data for feedstock potentials and costs, and for conversion technology performance and costs, as these generally make up the lion's share of overall costs. Table 5.4 provides which data sources were used over the course of the projects, and also how many categories of feedstock and conversion technologies were used.

Table 5.4 provides that the providers of the various datasets changed over the course of time, particularly on the side of feedstocks. Besides, the number of feedstock and technology categories changed. This is because of several drivers:

- In the number of feedstock categories, there are two countervailing drivers:
  - On the one hand, new feedstock types were introduced several times, broadening the scope of the resource base and providing more detail.
  - On the other hand, sometimes it was decided that some feedstock categories could be bundled, or specific categories contained negligible amounts of potential. This led to a reduction of categories, in the REFUEL and S2Biom model versions.
- In the number of conversion technologies, there were also two drivers:
  - In the REFUEL model version, various conversion technologies were aggregated because they essentially provided the same role.
  - The expansion of the output set of energy carriers (in the Biomass Futures project) led to a clear expansion of technologies, in order to cover the heat and power sectors, next to biofuels.

Generally, the number of feedstock types and conversion routes is always a compromise between level of detail and refinement and practical considerations such as model runtime and data collection efforts.

TABLE 5.4 Data Sources for Feedstocks and Conversion Technology Costs in BioTrans/RESolve-Biomass in the Various Projects

Project (Final Year) Model Name	Data Sources of:			
	Feedstock Costs, Potentials	No.[a]	Conversion Technology Performance and Costs	No.[b]
S2Biom (2016) RESolve-Biomass (van Stralen et al., 2016)	Crops and other: Biomass Policies (Alterra), verified against new S2Biom data (Un. Freiburg) Ex-EU imports: Biomass Policies (IINAS)	10 28 6	RES-H, -E and Biofuels: S2Biom (BTG) Bio-based chemicals: S2Biom (BTG, ECN)	77
Biomass Policies (2016) RESolve-Biomass (Uslu and van Stralen, 2016)	Crops and other: Biomass Policies (Alterra) Ex-EU imports: Biomass Policies (IINAS)	10 28 6	RES-H, -E: Biomass Policies (ECN) Biofuels: Biomass Policies (ECN)	72
Biomass Futures (2011) RESolve-Biomass (van Stralen et al., 2012; Uslu and van Stralen, 2012)	Crops and other: Biomass Futures (Alterra) Ex-EU imports: Biomass Futures (IIASA)	8 22 8	RES-H, -E: Biomass Futures (ECN) Biofuels: REFUEL and Biomass Futures (ECN)	46
Elobio (2010) BioTrans (Bole et al., 2010)	Crops: REFUEL (IIASA, Un. Utrecht) Other: REFUEL (ECN) Ex-EU imports: external	5 5 2	Biofuels: REFUEL (ECN)	12
REFUEL (2008) BioTrans (Deurwaarder et al., 2007)	Crops: REFUEL (IIASA, Un. Utrecht) Other: REFUEL (ECN) Ex-EU imports: external	5 5 2	Biofuels: REFUEL (ECN)	12
Viewls (2005) BioTrans (van Thuijl et al., 2005)	Crops: Viewls (Un. Utrecht) Other: Viewls (ECN) Ex-EU imports: –	5 11	Biofuels: Viewls (ECN)	24

[a]*Number of types of feedstock included in the model version. First number, crop types (including lignocellulosic crops); second number, other feedstock types; third number: import types.*
[b]*Number of conversion technologies in the model.*

5.4.2 Scenario Framework

Next to providing baseline or reference runs, almost all projects also worked with additional scenarios or policy cases (and in some projects the word "scenario" was applied to mere policy cases). Scenario analysis is a method of analyzing possible future situations, that highlight the central elements and draw attention to the key factors that will drive these developments. The bioenergy scenarios, in general, highlight how, under different plausible and consistent assumptions, the supply and demand of biomass may develop. One of the main objectives is to enable policymakers to systematically consider uncertainties inherent in bioenergy planning and to select strategies that are robust. The key characteristics of the various scenarios are indicated in Table 5.5.

TABLE 5.5 Main Scenario Assumptions of the Various Projects Reviewed

Project	Scenarios	Key Characteristics
S2Biom (van Stralen et al., 2016; Panoutsou, 2016)	High Central	• Faster technological learning and technology scale-up • Relatively lower fossil fuel prices, reflecting lower logistics costs
	Low Central	• Faster technological learning and technology scale-up • Relatively lower fossil fuel prices, reflecting lower logistics costs • Feedstock potential limitations reflecting sustainability, resource efficiency
	High decentral	• Slower technological learning and technology scale-up • Relatively higher fossil fuel prices indicating higher logistics costs
	Low decentral	• Slower technological learning and technology scale-up • Relatively higher fossil fuel prices indicating higher logistics costs • Feedstock potential limitations reflecting sustainability, resource efficiency
	Additional analyses	• Impacts of accelerated crop-based biofuels phaseout • Impacts of competition between energy and chemicals • Impacts of improved biomass mobilization or lower capital costs • Perspective for elimination of crop-based biofuels

(Continued)

TABLE 5.5 (Continued)

Project	Scenarios	Key Characteristics
Biomass Policies (Uslu and van Stralen, 2016)	Reference	Least-cost mix of bio-E/H and biofuels meeting NREAP targets
	Improved policies scenario	• Additional policies favoring resource efficiency and mobilization of biomass
Biomass Futures (Uslu et al., 2012; Uslu and van Stralen, 2012)	Reference	Least-cost mix of bio-E/H and biofuels meeting NREAP targets
	Alternative scenario	• Sustainability criteria applied to all energy sectors (HEF), not only on biofuels (as in reference), and also on biofuel imports • GHG threshold increased to 80% by 2030, taking into account ILUC emissions (reference: 50%, no ILUC)
	High biomass scenario	• Higher policy ambitions resulting in larger demand for biomass resources
Elobio (Bole et al., 2010)	Reference	Least-cost biofuels mix for a 15% biofuels share in 2030.
	Several policy cases	• Additional investment subsidies for advanced biofuels; combined with: • A tax break for advanced biofuels; or • A soft loan for advanced biofuels; or • Double counting of advanced biofuels toward a biofuels target
REFUEL (Londo et al., 2008; Londo, 2008)	Moderate case (reference)	Least-cost biofuels mix for a 15% biofuels share in 2030
	Several policy cases[a]	• Cases focusing on either GHG reduction, energy security, or innovation • Additional cases paying attention to either biodiversity or agriculture • A specific "REFUEL team" case
Viewls (Wakker et al., 2005)	Reference	Least-cost biofuels mix for a 20% biofuels share in 2030

[a]*These policy cases were built from coherent assumptions regarding (i) biofuels target height, (ii) the ambition levels for renewable heating and cooling (and corresponding biomass demand competing with biofuels), (iii) assumed levels of biofuels import, and presence or absence of (iv) a CO_2 price, (v) an energy crop premium, (vi) an investment subsidy, and (vii) a specific target for advanced biofuels.*

The overview indicates some frequently occurring features in scenarios or policy cases:

- In many projects, scenarios varied the availability and costs of feedstock. This clearly reflects the major uncertainties that are still associated with this domain. Feedstock potential limitations were usually motivated for uncertainty as such, or from sustainability and resource efficiency points of view. Sustainability considerations were once translated in specific additional thresholds for biofuels to be eligible.
- Policies for bioenergy and biofuels have always been developed from an amalgamate of motivations, including combating climate change, energy security, and building new economic activities. Next to these, considerations related to agriculture, forestry, and industry have played an important role. This means that many policy cases show the impact of differently balancing these motivations in the exact shaping of policies.
- The introduction of advanced biofuels (and the conditions under which this would occur) is also a returning point of attention. This is because of the better sustainability perspectives of these fuels, but also their high upfront costs. Several projects have done specific analyses on policies to mobilize advanced biofuels, and in almost all, resulting shares of conventional and advanced biofuels have been shown.

5.5 CONCLUSIONS AND RECOMMENDATIONS FOR FUTURE WORK

The review of model versions, data, and related research questions brings us to the following conclusions:

- The modeling framework of RESolve-Biomass is relatively flexible in terms of the questions that can be addressed by it. For each specific project, the model was used to address different research and policy questions, and with relatively minor modifications, useful results were generated.
- Also in terms of model coverage, the modeling framework appeared to be sufficiently flexible. New demand sectors, feedstock types, and technologies can be added relatively easily, and the same applies to the addition or deletion of countries (details not shown).
- A concern is that the modeling framework focuses on least-costs energy mixes, and energy units are the key denominator in the entire model. In principle, the model can also show CO_2 impacts, but full optimization on least costs per ton CO_2 is not possible and would require fundamental overhaul. With the increasing attention on CO_2 reduction as the key objective of energy policy, this is a point of attention.
- On the data side, the strongest changes in data have taken place in the feedstock part, with substantial changes in their cost−supply curves over

the course of the projects. It seems that conversion technology data are less disputed, although this can also be caused by the fact that one party (ECN) was always involved in the provision of these data.

- A key factor in the relevance of model outcomes is the shaping of scenarios and policy cases. A reference scenario as such may be useful but will probably never reflect the real future: after all, models are usually wrong in predicting the future. Comparing various scenarios or policy cases is often a much more useful exercise, because such differences may bring more concrete insights in the impact of specific uncertainties or policy measures. In this context, the construction of scenarios and policy cases and their translation into (differences in) model parameters is a critical part of the research work.

The key domains in which the modeling could be further improved strongly depend on the kind of future questions that would need to be answered by the model results. Some suggestions can be done:

- The modeling of bio-based chemicals, although adding a significant amount of new technologies, still does not cover the full diversity of chains and technologies at stake for this sector. This could be further detailed, also in order to better estimate economic competitiveness of bio-based chemicals, and their sensitivity to specific policy incentives.
- In the use of biomass for energy and chemicals, biorefinery concepts are generally believed to become more and more important: industrial complexes in which various types of feedstock (biomass and possibly also fossil resources) are converted into an array of products and energy carriers. While RESolve-Biomass is currently able to include relatively simplified biorefineries (e.g., coproduction of biofuels, heat and power, and CHP), further details on biorefineries would add value. This could also include making biorefinery outputs flexible in terms of the relative shares of different products and energy carriers.
- The time horizon of the model is still 2030. When it was developed in the early 2000s, that was sufficiently far away. However, as fundamental changes in our energy economy are expected to take place also after 2030, a time horizon expansion is clearly desired. This will require significant addition of new data, both on the demand sectors, feedstock potentials, and costs, and conversion technology performances.
- A key difference between RESolve-Biomass and complete IAMs is that the competition between bio-based energy carriers and other renewable options (e.g., bio-based electricity vs wind and solar power) is relatively simplified. Also, the model used for this, the sister model RESolve-E, has been updated less frequently than RESolve-Biomass. Especially in the context of meeting overall renewables targets, it may be important to improve this point.

- While the model does include CO_2 information, it does not have a specific link with land use models and other tools that are spatially more explicit. The most detailed level on which the model can show results is NUTS-2. It may be argued that such links should be improved. On the other hand, issues related to land use change may require a fundamentally different way of modeling (think of general equilibrium models, for example), and RESolve-Biomass will have other questions it is better able to answer.

REFERENCES

Bole, T., Londo, M., van Stralen, J., Uslu, A., 2010. Overcoming the Initial Investment Hurdle for Advanced Biofuels: An Analysis of Biofuel-Related Risks and Their Impact on Project Financing; Report D7.1 of ELOBIO. ECN, Amsterdam.

de Jong, S., Hoefnagels, R., van Stralen, J., Londo, M., Slade, R., Faaij, A., et al., 2017. Renewable Jet Fuel in the European Union—Scenarios and Preconditions for Renewable Jet Fuel Deployment Towards 2030. Utrecht University, Utrecht.

de Wit, M.P., Junginger, H.M., Lensink, S.M., Londo, M., Faaij, A.P.C., 2010. Competition between biofuels: modeling technological learning and cost reductions over time. Biomass Energ. 34 (2), 203–217.

Deurwaarder, E.P., Lensink, S.M., Londo, H.M., 2007. BioTrans Biofuels Data; Appendix to "Use of BioTrans in Refuel"; Functional and Technical Description. Refuel Deliverable D10b. ECN, Amsterdam.

Kraan, C., 2015. Market Analysis of Heat, Electricity and Biofuels. D7.2a of S2Biom. ECN, Amsterdam.

Lensink, S.M., Londo, M., Deurwaarder, E.P., 2007. Use of BioTrans in REFUEL Functional and Technical Description; Report of REFUEL WP4. ECN, Amsterdam.

Liebmann, L., Resch, G., Welisch, M., 2015. Scenario Database "Long-Term Prospects for RES Cooperation With Third Countries Beyond 2020". D6.3 of the BETTER Project. EEG/TU Wien, Vienna.

Londo, H.M., Lensink, S.M., Deurwaarder, E.P., Wakker, A., de Wit, M.P., Junginger, H.M., et al., 2008. Biofuels Development in the EU27 + Until 2030; Full-Chain Cost Assessment and Implications of Policy Options. REFUEL WP4 Final Report. ECN, Amsterdam.

Londo, M., 2008. Eyes on the track, Mind on the horizon; From inconvenient rapeseed to clean wood: a European road map for biofuels. REFUEL final report. ECN, Amsterdam.

Loulou, R., Goldstein, G., Noble, K., 2004. Documentation for the MARKAL Family of Models. Energy Technology Systems Analysis Programme (ETSAP), Paris.

Loulou, R., Kanudia, A., Lettila, A., Remme, U., 2016. Documentation for the TIMES Model—Part I. Energy Technology Systems Analysis Programme (ETSAP), Paris.

Mozaffarian, M., Londo, M., Pantaleo, M., Kraan, C., 2016. Market Analysis for Lignocellulosic Biomass as Feedstock for Bioenergy, Biobased Chemicals & Materials in Europe; A Quantitative Estimate of Biomass Demand in 2020 and 2030. S2Biom Deliverable 7.2. ECN, Amsterdam.

Panoutsou, C., 2016. Extensive Description of Scenarios; S2Biom Deliverable 7.1. Imperial College, London.

Pantaleo, M., 2015. Market analysis of sugar-lignin platforms. S2Biom Delilverable 7.2b. Imperial College, London.

Uslu, A., van Stralen, J., 2012. Biomass Futures Scenario Set-up and the Methodology for Analysis. ECN, Amsterdam.

Uslu, A., van Stralen, J., 2012. Biomass Futures Scenario Setup and the Methodology for Analysis. Deliverable 5.2. ECN, Amsterdam.

Uslu, A., van Stralen, J., 2016. Effects of Policy Framework in the Bioenergy Market; Uncertain Policy Framework, Future of Advanced Biofuels, Biomass Co-firing Issue and the Competition Dilemma. Biomass Policies Deliverable 4.2 (b). ECN, Amsterdam.

Uslu, A., van Stralen, J., Beurskens, L., dalla Longa, F., 2012. Use of Sustainable Biomass to Produce Electricity, Heat and Transport Fuels in EU27—A Model-Based Analysis of Biomass Use for 2020 and 2030. ECN, The Netherlands.

van Stralen, J., dalla Longa, F., Beurskens, L., Uslu, A., 2012. Functional Description of the RESolve Model Kit and the Biomass Allocation. ECN, Amsterdam.

van Stralen, J., dalla Longa, F., Beurskens, L., Uslu, A., 2012. Functional Description of Biomass Allocation Within the RESolve Model Kit. ECN, The Netherlands, Report No: Deliverable D5.1 of Biomass Futures.

van Stralen, J., Kraan, C., Uslu, A., Londo, M., Mozaffarian, M., 2016. Integrated Assessment of Biomass Supply Chains and Conversion Routes Under Different Scenarios; Deliverable 7.3 of the S2Biom Project. ECN, Amsterdam.

van Stralen, J., Uslu, A., Panoutsou, C., 2016. Future Policy Formation Towards Resource Efficiency; Effects of Policy Measures on Selected Priority Value Chains in Eleven Member States. Biomass Policies Deliverable 4.2 (a). ECN, Amsterdam.

van Thuijl, E., Wakker, A., Beurskens, L.W.M., Deurwaarder, E.P., Egging, R., 2005. VIEWLS Modelling and Analysis; Technology Data for Biofuel Production Chains. ECN, Amsterdam.

van Tilburg, X., Egging, R., Londo, M., 2005. BIOTRANS Functional and Technical Description; Report of VIEWLS WP5, Modelling Studies. ECN, Amsterdam.

Wakker, A., Egging, R., van Thuijl, E., van Tilburg, X., Deurwaarder, E.P., de Lange, T., et al., 2005. Biofuel and Bioenergy Implementation Scenarios; Final Report of VIEWLS WP5, Modelling Studies. ECN, Amsterdam.

Chapter 6

Assessing Lignocellulosic Biomass Potentials From Forests and Industry

Marcus Lindner, Matthias G. Dees, Perttu Anttila, Pieter J. Verkerk, Joanne Fitzgerald, Pawan Datta, Branko Glavonjic, Robert Prinz and Sergey Zudin

6.1 INTRODUCTION

Wood or biomass from forests and forest industry has by far the largest share of nonfood biomass feedstock supply in Europe. It is used as a resource to produce materials (e.g., furniture, construction materials, paper, and packaging products), chemicals, as well as renewable energy. Many resource assessments have been carried out over the last decade to quantify biomass from forests and forest industry (e.g., UNECE and FAO, 2011; Verkerk et al., 2011; Kindermann et al., 2013; Mola-Yudego et al., 2017).

The analysis of total lignocellulosic biomass potentials includes the analysis of potentials in forests, potentials outside forests, and potentials in short rotation coppice for fast-growing wood species, which can be grown on forest land or unused agricultural land. Moreover, woody biomass can consist of fresh biomass production (stemwood from final fellings, thinnings, or landscape maintenance), primary residues from forests (logging residues, stumps), secondary forest biomass residues in the wood-processing industry (e.g., sawdust, woodchips, and black liquor), and postconsumer wood (e.g., demolition wood and discarded wood products). Here we focus on forest biomass and primary forest residues (Section 6.2) and secondary wood residues (Section 6.4). Short rotation coppice in forests is also included, whereas coppices on agricultural land are covered under the chapter on agricultural potentials. Postconsumer wood residues are addressed in the waste biomass chapter.

Modeling and Optimization of Biomass Supply Chains.
DOI: http://dx.doi.org/10.1016/B978-0-12-812303-4.00006-9

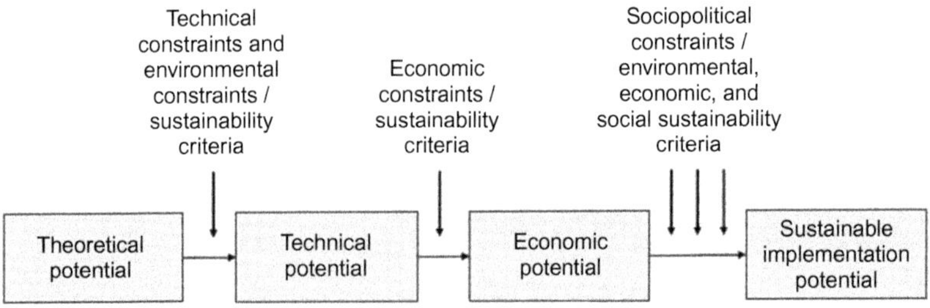

FIGURE 6.1 The integration of sustainability criteria in biomass potential assessments (Vis and Dees, 2011).

Assessments of biomass potentials are often difficult to compare, because there are many different aspects considered. Fig. 6.1 shows one way of separating biomass potentials into (1) theoretical potential, (2) technical potential, (3) economic potential, and (4) implementation potential (Vis and Dees, 2011). The theoretical potential is the overall maximum amount of terrestrial biomass, which can be considered theoretically available within fundamental biophysical limits. Technical, economic, and implementation potentials are subject to many assumptions and often difficult to compare across studies. The S2Biom project has specified its own definitions following the same principals and nomenclature across all sectors, forestry, agriculture, nonfood crops, and waste, which were applied in this chapter to quantify European lignocellulosic biomass potentials with state-of-the art methods and data from different biomass categories.

Within S2Biom biomass potentials were estimated for the following types:

- The *technical potential* is the absolute maximum amount of lignocellulosic biomass potentially available assuming the absolute minimum of technical constraints.
- The *base potential* represents the potential most closely aligned to current guidelines of sustainable forest management. This also covers legal restrictions such as restrictions from management plans in protected areas, e.g., Natura 2000.
- The *high potential* is a potential with less constraints compared to the base potential, assuming a strong focus on the use of wood for producing energy. It includes a strong mechanization of harvesting across Europe. Biomass harvesting guidelines are less restrictive, e.g., stumps are included in this potential for all S2Biom countries.
- In addition, eight *user-defined potentials* can be derived from the base potential, varying in terms of type and number of considerations per biomass assortment.

The base and high potentials include the constraints considered for the technical potential, but take additional constraints into account. The user-defined potentials consist of four options. Each of these options allow the user to evaluate the effect of changes to protected forest area, harvesting practice, or technical and environmental constraints, and to evaluate the potential availability of biomass under differing policy environments.

Particular focus in the S2Biom biomass assessment was placed on the quantification of cost−supply potentials of forest biomass, which were lacking from previous European scale assessments. Besides the European Union (EU) scale analysis, dedicated efforts were also made to study biomass potentials in the southeast neighboring region of the EU.

6.2 PRIMARY FOREST BIOMASS POTENTIALS

6.2.1 Definitions

Terms and definitions related to forest biomass are presented in Table 6.1.

6.2.2 Methodology for Assessing Primary Forest Biomass

The estimation of primary woody biomass potentials from forests (i.e., potentials for wood production and primary forest residues) was estimated following the approach previously described and applied by UNECE and FAO (2011), Verkerk et al. (2011), and Verkerk (2015) for the period from 2012 to 2030. First, we applied the large-scale European Forest Information SCENario model (EFISCEN; versions 3.1.3 and 4.0) (Sallnäs, 1990; Schelhaas et al., 2007; Verkerk et al., 2016a) to estimate the theoretical potential of forest biomass supply in Europe based on detailed forest inventory data, taking into account increment, the age structure, and stocking level of the forests. Second, multiple environmental and technical constraints were defined and quantified that reduce the amount of extractable biomass from forests for different biomass potential types. Third, the theoretical potentials from the first step were combined with the constraints for the biomass potential types. In S2Biom, spatially disaggregated woody biomass potentials were estimated based on a modified typology of potentials, using updated forest inventory data and improved sets of constraints. The geographic scope of the assessment included 37 S2Biom countries in Europe. Biomass potentials were calculated in cubic meters and converted to tonnes dry matter utilizing generic conversion factors (0.542 t/m^3 for nonconifers and 0.415 t/m^3 for conifers).

The structure of the European forest resources in 2010 was estimated using data from recent national forest inventories. In case only forest inventories from older inventories were available, the EFISCEN model was used to estimate forest resource structure in 2010, using historical roundwood production (FAOSTAT) converted to overbark volumes (assuming a bark fraction of 12%). To estimate the biomass potentials from forests, the EFISCEN

TABLE 6.1 Forest Biomass Definitions

Biomass Type	Definition	Source (Reference)
Stemwood	Part of tree stem with the branches and top removed, with a length of more than 100 cm.	EN ISO 16559:2014
Logging residues	*Woody biomass residues* created during wood harvesting. *Logging residues* include branches and tree tops that can be salvaged when fresh or after seasoning.	EN ISO 16559:2014
Stump	Part of the tree stem below the felling cut. In total-tree utilization the root system is included in the stump.	EN ISO 16559:2014
Other Relevant Definitions Related to Forest Biomass		
Standing volume	Volume of standing trees, living or dead, above-stump measured overbark to top (0 cm). Includes all trees with diameter over 0 cm (d.b.h.).	UNECE/FAO, http://www.unece.org/timber/fra/definit.htm
	Includes: Tops of stems, large branches; dead trees lying on the ground that can still be used for fiber or fuel.	
	Excludes: Small branches, twigs, and foliage.	
Growing stock	The living tree component of the standing volume.	UNECE/FAO, http://www.unece.org/timber/fra/definit.htm
Net annual increment	Average annual volume over the given reference period of gross increment less that of natural losses on all trees to a minimum diameter of 0 cm (d.b.h.).	UNECE/FAO, http://www.unece.org/timber/fra/definit.htm

model was subsequently used to iteratively assess the theoretical harvest potential of stemwood for the period 2010–30 for every 5-year time-step, as well as the associated potential from logging residues and stumps, from thinning and final fellings separately (see Verkerk et al., 2011, for details).

Due to absence of the required forest inventory data, EFISCEN could not be applied for Cyprus, Greece, Montenegro, Former Yugoslav Republic of Macedonia (FYROM), Bosnia and Herzegovina, Kosovo, Malta, and Serbia. Instead, woody biomass potentials for these countries (except for Malta) were estimated as described in the biomass handbook developed in Biomass

Energy Europe (BEE) (Vis and Dees, 2011); this approach assumes that the theoretical stemwood harvest potential was based on the net annual increment corrected for harvesting losses. We used aggregated data on forest area and net annual increment from Forest Europe (2015) or—in case of missing data—on Forest Europe (2011). For Greece we used net annual increment as reported by Meliadis et al. (2010). We used biomass allocation functions from Teobaldelli et al. (2009) and estimated stump biomass based on data by Asikainen et al. (2008). No data were available for Malta.

Constraints on biomass supply were identified from literature and were quantified separately by type of biomass (i.e., stemwood, logging residues, and stumps) and by type of felling activity (i.e., thinnings and final felling) for the different biomass potential types. An example of the constraints is given in Table 6.2. The complete resource assessment methodology with all assumptions is documented by Dees et al. (2017a).

Spatially explicit data sets were collected for each of the constraints and they were combined with the relevant constraint values for the different potential types. A raster layer was created for each constraint with a resolution of $1 \times 1\ km^2$. Finally, all relevant layers were combined and the lowest, permitted extraction rate according to each potential type defined for each pixel.

The EFISCEN disaggregation tool (Verkerk et al., 2016b) was used to disaggregate the estimated biomass potentials to the 1 km grid level using tree species distribution maps (Brus et al., 2012). The disaggregated woody biomass potentials were then multiplied with the respective constraint map to calculate the constrained potential at grid level. The resulting rasters were then again reaggregated to NUTS3 regions.

6.2.3 Results

Primary forest biomass potentials from forests in the EU and nine neighboring countries in 2012 were estimated at 379 million tonnes dry matter/year overbark (or 817 million m^3/year overbark), according to the base potential. The potential was projected to be rather stable over time with only a slight decrease to 370 million tonnes dry matter/year overbark by 2030. About 88% of the total biomass potential was in stems, while logging residues and stumps represented 12%; 45% of the total potential supply of the study area was concentrated in only four countries (Sweden, Germany, France, and Finland), mainly because of the extent of their forest areas (Fig. 6.2).

The calculated biomass availability is strongly affected by the type of potential and constraints considered (Fig. 6.3). The high potential with strong mechanization and weaker environmental constraints results in 17% higher biomass potentials compared to the base potential, mainly due to an increased potential of logging residues and stumps. On the other hand, more strict biodiversity protection constraints reduce the potentials by 11% to

TABLE 6.2 Maximum Extraction Rates for Extracting Logging Residues From Final Fellings Due to Environmental and Technical Constraints

Type of Constraint	Technical Potential	High Potential	Base Potential
Site productivity	Not a constraining factor	Not a constraining factor	35% extraction rate on poor soils (Acrisol, Podzoluvisol, Histosol, Podzol, Arenosol, Planosol, Xerosol); not a constraining factor on other soils
Soil and water protection: ruggedness	Not a constraining factor	70% on slopes up to TRI[1] highly rugged; 0% over TRI highly rugged	70% on slopes up to TRI[1] moderately rugged; 0% over moderately rugged
Soil and water protection: soil depth	Not a constraining factor	Not a constraining factor	0% on Rendzina, Lithosol, and Ranker (very low soil depth)
Soil and water protection: soil surface texture	Not a constraining factor	Not a constraining factor	0% on peatlands (Histosols)
Soil and water protection: soil compaction risk	Not a constraining factor	Not a constraining factor	0% on soils with very high compaction risk; 25% on soils with high compaction risk; not a constraining factor on other soils
Biodiversity: protected forest areas	Not a constraining factor	0%; not a constraining factor in areas with high or very high fire risk	0%; not a constraining factor in areas with high or very high fire risk
Recovery rate	70% up to TRI[1] highly rugged; 0% over TRI highly rugged	70% on slopes up to TRI highly rugged; 0% over TRI highly rugged	70% on slopes up to TRI moderately rugged; 0% over moderately rugged
Soil-bearing capacity	0% on Histosols, Fluvisols, Gleysols, and Andosols, not a constraint in Finland and Sweden	0% on Histosols, Fluvisols, Gleysols, and Andosols, not a constraint in Finland and Sweden	0% on Histosols, Fluvisols, Gleysols, and Andosols

[1]Terrain ruggedness index

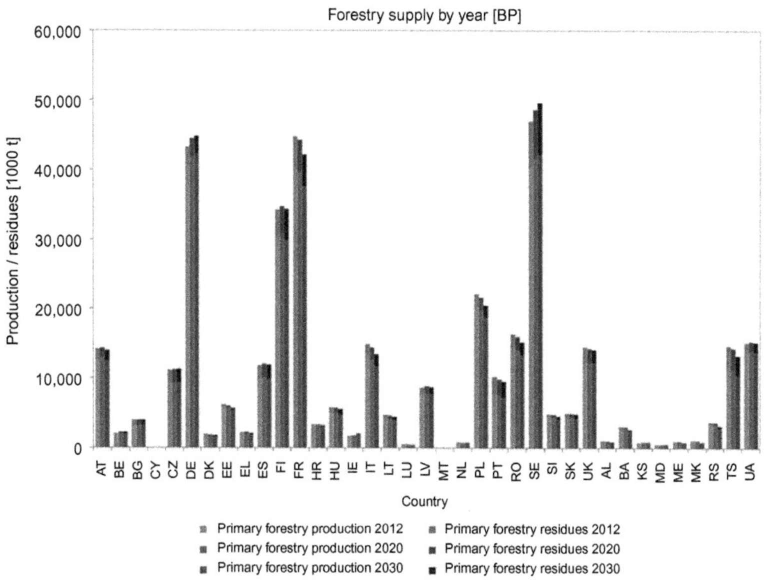

FIGURE 6.2 Total forestry potentials by country for the base potential in 2012, 2020, and 2030.

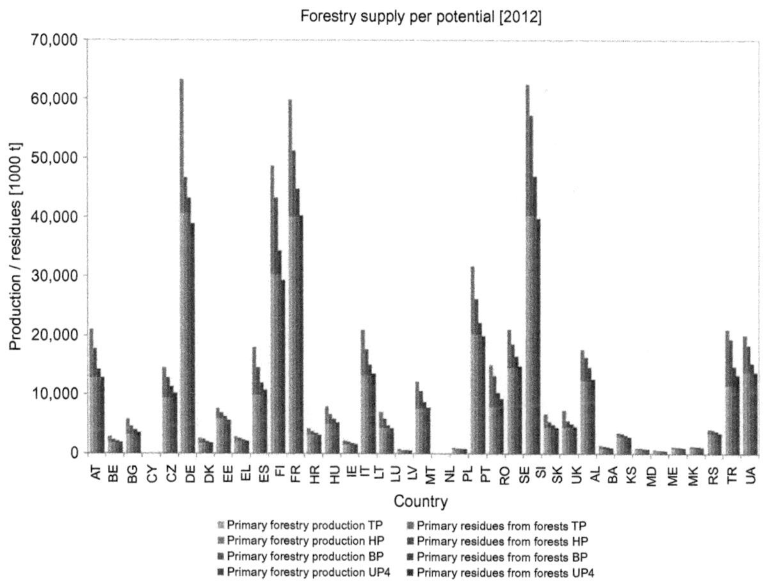

FIGURE 6.3 Total forestry potentials by country for the base potential (BP), high potential (HP), technical potential (TP), and a user-defined potential with enhanced biodiversity protection (UP4) in 2012.

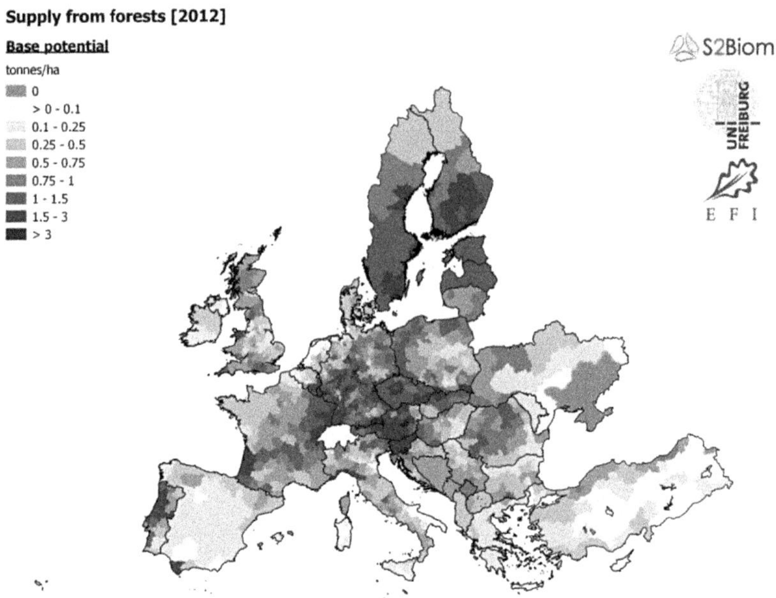

FIGURE 6.4 Total primary forest biomass supply potential (biomass production and primary residues from forests) per hectare of land at NUTS3 level for the base potential in 2012.

336 million tonnes dry matter/year overbark. The regional availability of biomass varies strongly across Europe. The largest potentials can be found in regions with high forest cover, productive tree species, and in mountain areas with low management intensity (Fig. 6.4).

6.3 SUPPLY COST OF PRIMARY FOREST BIOMASS

6.3.1 Cost Assessment Methodology

The estimation of harvesting and comminution costs is following the approach presented earlier by Ranta (2002, 2005), Ilavský et al. (2007), Anttila et al. (2011), and Laitila et al. (2015). Wood production costs are not considered in the cost estimates. The general workflow is illustrated in Fig. 6.5. The methodology can be divided into two main components: (1) the estimation of hourly machine costs, and (2) the estimation of machine productivity. All the cost estimations pertain to current cost level (year 2012).

To enable better comparison of costs between regions, supply chains were standardized. The dominant supply chain of stemwood to energy production in Europe is the chain based on roadside chipping (Díaz-Yáñez et al., 2013; Junginger et al., 2005; Eriksson et al., 2013; Holzleitner et al., 2013; Rottensteiner et al., 2013; Routa et al., 2013; Wolfsmayr and Rauch, 2014). In the chain felling and bunching are carried out by a harvester,

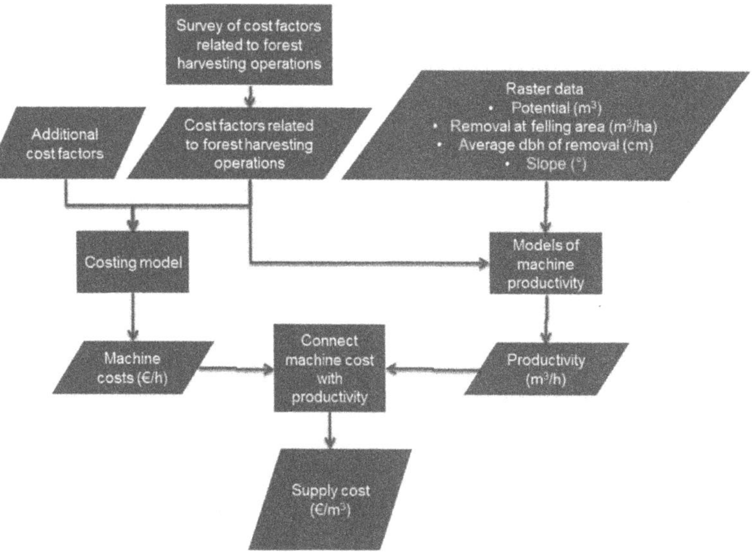

FIGURE 6.5 General workflow of the forest biomass cost calculations.

off-road transport by a forwarder, and chipping by a mobile chipper. For logging residues, the chain is otherwise similar except for the missing felling phase. Instead, piling of logging residues by the harvester is considered to belong to logging residue supply chain. Stumps are extracted by an excavator, forwarded to roadside and crushed by a mobile grinder.

6.3.1.1 Machine Costs Calculations

The costing model developed for the Cost Action FP0902 was utilized in machine cost calculations (Ackerman et al., 2014). For this study, the input of the costing model consisted of both machine-level and country-level data. The degrees of machine utilization were based on the study by Laitila et al. (2010). Fuel consumption data has been collected by Laitila et al. (2012) except for the grinder, for which the consumption has been measured by Eliasson et al. (2012). Oil and lubricant costs as well as maintenance and repair costs have been assessed by Laitila and Väätäinen (2011) and Nurminen et al. (2009). For all the machines 1700 productive hours per annum and 12,000 in total (Laitila et al., 2010) were assumed. For the other input parameters, the default values of the costing model were assumed.

Price level indices of 2012 were used to calculate machine prices for each country after the country-specific indices (EUROSTAT, 2015a) based on the average Finnish machine purchasing prices. The price level indices for Ukraine and Moldova were assumed to be the same as in Romania, for Kosovo the Serbian values were assumed.

The rest of the country-level data consisted of fuel price, labor costs, and interest rate. The fuel costs were taken from oil bulletin (DG Energy, 2015), prices of automotive diesel in 2012 with taxes reduced by the VAT in 2012. The fuel costs for Albania, Bosnia and Herzegovina, Croatia, Macedonia, Montenegro, Serbia, Kosovo, Turkey, Ukraine, and Moldova were calculated by multiplying the pump fuel price available for all countries (obtained October 30, 2015, from http://www.fuel-prices-europe.info/) with the ratio between the average fuel price of available countries without VAT and the pump fuel price on October 30, 2015.

The data source for the labor costs and their definition was EUROSTAT (2015b). As no data for the exact work categories of this study were available, labor costs of class "industry—except construction" were used for felling, stump extraction, chipping and grinding, and costs of class "transportation and storage" for forwarding. For Malta the missing transportation value was calculated by multiplying the industry value by the average ratio of costs for "transportation and storage" and the costs "industry—except construction." For countries where the data were missing, the numbers of neighboring countries were used. Survey data (Dees et al., 2015) was used in calculation only for Ukraine, for Moldova the numbers of Ukraine were used.

As a reference interest rate for forest machine investments, MFI indicators have been used (European Central Bank, 2013). The values to loan size "over an amount of €250,000 and up to €1 million" and the category "floating rate and up to 3 months initial rate fixation" with the average of all the months in 2012 have been used. For all countries the Euro zone values were applied.

6.3.1.2 *Productivity Calculations*

Productivities of machines and the dependences of the productivities on the operating environment have been formulated in productivity models based on field measurements. In order to enable the comparability between countries, only one model was selected for each biomass category and machine (Table 6.3). The productivities (m^3/h) and further unit costs (€/m^3) were estimated at NUTS3 (or corresponding regions defined in the project) level.

In addition to the properties of forest machines, a large part of their productivity can be explained by the properties of the operating environment. Information on the properties of forest stands to be harvested were obtained in raster format based on calculations carried out in Section 6.2 with the EFISCEN model. The information included the intensity of harvesting (m^3/ha) for final fellings and thinnings separately for conifers and non-conifers as well as the average diameter (cm) of removed trees for the same classes in a 1 km $\times$ 1 km^2 grid. The grid data were aggregated to NUTS3 level to be used as input parameters for productivity models.

TABLE 6.3 The Applied Productivity Models by Biomass Categories (Broadleaf and Conifer Trees)

Biomass Category	Mechanized Felling and Bunching	Integrated Bunching While Felling With a Harvester	Forwarding	Chipping	Stump Lifting	Crushing
Stemwood from final fellings	Kuitto et al. (1994)	N/A	Kuitto et al. (1994)	Laitila (2006)	N/A	N/A
Stemwood from thinnings	Kuitto et al. (1994)	N/A	Kuitto et al. (1994)	Laitila (2006)	N/A	N/A
Logging residues from final fellings	N/A	Laitila (2006)	Ranta (2002)	Laitila (2006)	N/A	N/A
Stumps from final fellings	N/A	N/A	Laitila (2010)	N/A	Laitila et al. (2008)	Nuutinen et al. (2014)

In case harvesting intensity or diameter data were missing, a calculation using average values of near-neighbor countries or regions was performed. In some rare cases the estimated values for harvesting intensity or diameter were outside the range of data values used in modeling productivity resulting in unrealistic cost estimates. In these cases costs could not be estimated.

The effect of slope was considered by multiplying the time consumption of forwarding by a driving speed factor determined by the steepness of slope (Ilavský et al., 2007). The slope values were calculated by using the Digital Elevation Model over Europe from the GMES RDA project in a spatial resolution of 25 m that was derived from the original posting of 1 arc sec ($\sim$30 m) with an overall vertical accuracy of 2.9 m RMSE (root mean square error) (DHI GRAS 2014). Other factors describing the operating environment were acquired from Dees et al. (2015) or literature.

The intensity of harvesting or average diameter may not be directly usable as explanatory variables of the productivity model. In order to estimate such variables (e.g. tree biomass or volume), the countries were divided into four groups: NEU (Northern Europe), CEU (Central Europe), EEU (Eastern Europe), and SEU (Southern Europe).

Representative species were defined for each country group: one species for broadleaves, one for conifers (except for Southern Europe). The biomass and stem volume functions applied for the four country categories are given in Table 6.4.

The process of felling and bunching comprises of moving in the felling area and processing the trees (Kuitto et al., 1994). The time consumption of moving depends on the number of removed trees per hectare, estimated based on the harvesting intensity and average stem volume (which explains tree-processing time). Finally, the productivity of felling and bunching was calculated as a function of the two above-mentioned and corrected by factor of 1.3, which was assumed to be the ratio between gross effective and effective time (Laitila et al., 2010).

The process of forwarding consists of loading the forwarder, driving during loading, driving when empty, driving when loading and unloading the forwarder (Kuitto et al., 1994). Time consumption of loading and driving during loading depend on harvesting intensity and the load size of the forwarder (Table 6.5), the one of driving on forwarding distance (Dees et al., 2015) and the one of unloading on the load size. The gross effective/effective time ratio was set at 1.2 (Laitila et al., 2010).

The productivity of chipping per effective machine hour was assumed to be 85 loose cubic meters (Laitila, 2006). The degree of machine utilization was set at 65% (Laitila et al., 2010).

The costs of wood production were not considered in this study.

With logging residues from final fellings, no direct cost is incurred for felling. However, for logging residues to be forwarded efficiently, they have to be bunched by the harvester during felling. A small compensation for bunching of 0.3 €/m^3 was used (Laitila, 2006). The models of Ranta (2002)

TABLE 6.4 The Biomass and Volume Functions Applied in the Study

Biomass Category	NEU	CEU	EEU	SEU
Stemwood from final fellings originating from broadleaf trees	Repola (2008)	Pellinen (1986)	Pellinen (1986)	Brandini and Tabacchi (1996)
Stemwood from final fellings originating from conifer trees	Repola (2009)	Černý (1990)	Černý (1990)	Corona and Ferrara (1987)
Stemwood from thinnings originating from broadleaf trees	Repola (2008)	Pellinen (1986)	Pellinen (1986)	Brandini and Tabacchi (1996)
Stemwood from thinnings originating from conifer trees	Repola (2009)	Černý (1990)	Černý (1990)	Corona and Ferrara (1987)
Logging residues from final fellings originating from broadleaf trees	Repola (2008)	Pretzsch (2000)	Pretzsch (2000)	Brandini and Tabacchi (1996)
Logging residues from final fellings originating from conifer trees	Repola (2009)	Fiedler (1986)	Černý (1990)	Menguzzato and Tabacchi (1988)
Stumps from final fellings originating from broadleaf trees	Repola (2008)	N/A	N/A	N/A
Stumps from final fellings originating from conifer trees	Repola (2009)	N/A	N/A	N/A

were used for time consumption for forwarding. Chipping productivity was set at 70 loose cubic meters (Laitila, 2006).

Stump extraction time was divided to excavator moving time and processing time (Laitila et al., 2008). The moving time was determined by the intensity of harvesting. The processing time is explained by stump diameter. The gross effective/effective time ratio for an excavator was set at 1.1 (Laitila et al., 2010). The time consumption of forwarding was estimated with the models by Laitila et al. (2008). Grinding productivity was set at the level of 36 m^3/h observed by Nuutinen et al. (2014).

Further details on the cost assessment methodology are documented in Dees et al. (2017a).

6.3.2 Results

Generally, the supply costs in Eastern Europe tend to be lower than elsewhere in the study area (see Fig. 6.6 as an example). Further work would be needed to thoroughly analyze which of the explanatory variables are the most influential to the costs. However, the lower costs in Eastern Europe

TABLE 6.5 Forwarder Load Capacity in Solid Cubic Meters for the Different Biomass Categories

	m^3	Source
Stemwood from final fellings originating from broadleaf trees	10.24	Kuitto et al. (1994)
Stemwood from final fellings originating from conifer trees	12.8	Kuitto et al. (1994)
Stemwood from thinnings originating from broadleaf trees	8	Kuitto et al. (1994)
Stemwood from thinnings originating from conifer trees	10	Kuitto et al. (1994)
Logging residues from final fellings originating from broadleaf trees	7.8	Laitila et al. (2010)
Logging residues from final fellings originating from conifer trees	7.8	Laitila et al. (2010)
Stumps from final fellings originating from broadleaf trees	8.6	Laitila et al. (2010)
Stumps from final fellings originating from conifer trees	8.6	Laitila et al. (2010)

can—at least to a certain extent—be explained by lower investment, fuel, and especially labor costs.

In Western Europe Denmark, Italy, and Ireland stand out from the neighboring countries with respect to the supply cost of stemwood from final fellings. This might be due to the small tree size, relatively long forwarding distances, and—in Italy—steep terrain. The supply costs of logging residues from final fellings seem to be especially high in Romania, Greece, and Italy, possibly owing to the long forwarding distances. Furthermore, small tree size explains the exceptionally high supply cost of nonconifer stemwood from thinnings in Ukraine and in Latvia.

Fig. 6.7 compares selected cost−supply curves from Austria, Finland, Italy, and Poland and clearly shows the different levels of costs and supply in the four countries. Although the curves are highly averaged, they underline the varying operating environments across Europe. In Poland the supply costs are relatively low, while the potential of the whole country is quite high. On the contrary, in Italy the potential remains low in spite of high supply costs. There is also extremely wide range in cost among the Italian regions. The costs in Austria and Finland fell between the costs in Poland and Italy. In Finland the costs are slightly higher than in Austria, but the potential is considerably bigger.

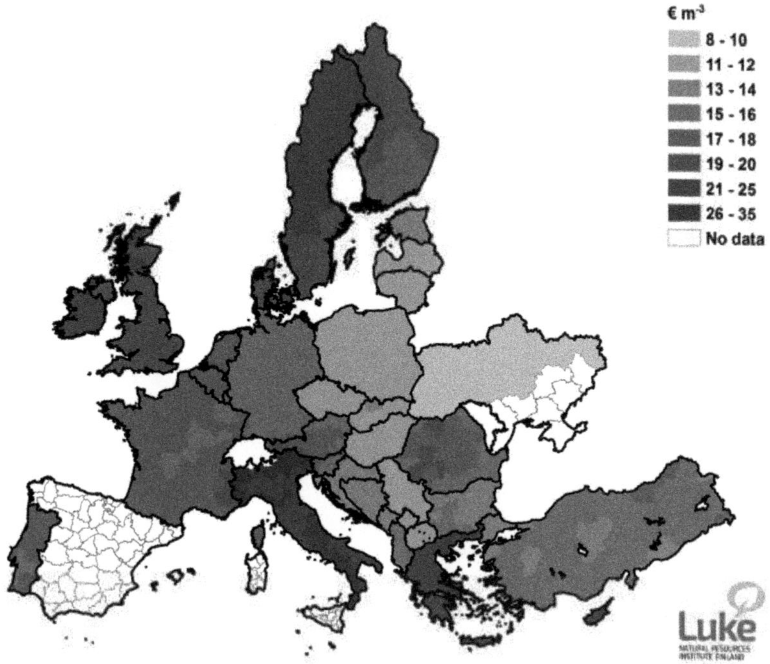

FIGURE 6.6 Regional variation in supply costs of logging residues from final fellings of conifer trees.

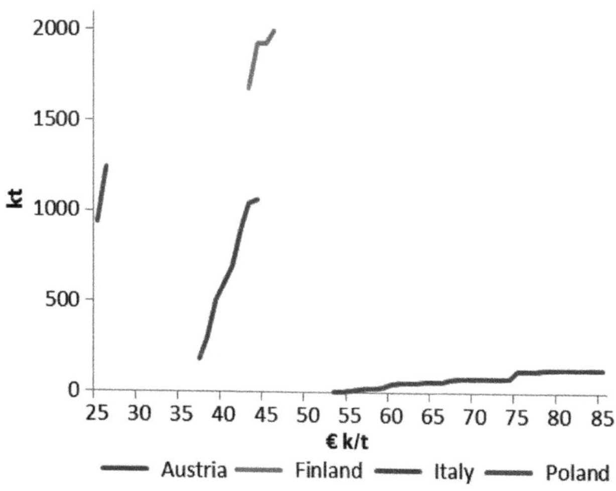

FIGURE 6.7 Cost–supply curves of logging residues from final fellings of conifer trees in four countries based on NUTS3-level data.

6.4 SECONDARY FORESTRY RESIDUES

6.4.1 Definitions

The secondary forestry residues (wood-processing residues) can be grouped into resource segments following main industry branches: sawmill by-products, black liquor (residues of the pulp industry), and other industrial wood residues. The residues include various types of biomass originating during industrial processing of timber, e.g., sawdust and cutter chips, bark, slabs, lump wood residues, and black liquor. Secondary forestry residues are a spatially concentrated resource—large amounts of the residues can be available from a single factory. In contrast to other types of forest woody biomass, all secondary forestry residues are technically accessible. Moreover, secondary solid forestry residues usually have lower moisture content than primary forestry residues and, as a result, a higher net calorific value.

In Table 6.6 basic terms relevant in the context of the secondary forest residues are defined.

Following the generic general potential definition of S2Biom, technical and base potentials are identical since neither accessibility constraints nor other constraints have been applied. Still user-defined potentials have been determined to allow insight in the availability of different viewpoints both as such, but as well to allow specific combinations with the potentials from forests.

User-defined option 1 residues from saw mill industry that are utilized by other wood industry sectors are not accounted as saw mill industry residues, since they are not available for energy or other options of utilization.

User-defined option 2 considers only residues from sawmills and veneer and plywood industry. These industry sectors cover the clear majority of utilization of high-quality assortment wood, whereas the other sectors (board, paper, energy, chemical use) can be seen as relatively direct competitive utilizations for assortments of comparable value.

6.4.2 Methodology for Secondary Forestry Residue Assessment

For the estimation of the potentials of secondary forestry residues, methodological approaches were developed within the project BEE (Vis and Dees, 2011) and the EUwood project (Saal, 2010a,b). The method adopted in S2Biom builds on the latter methodological concepts.

The secondary forestry residues include the following categories of residues:

- Sawmill residues, separated in:
  - sawdust conifer,
  - sawdust nonconifer,

TABLE 6.6 Secondary Forestry Residues Definitions

Biomass By-product	A secondary product, which is made incidentally during the production of something else (e.g., sawdust when sawing timber).	EN ISO 16559:2014
Bark	Organic cellular tissue, which is formed by taller plants (trees, bushes) on the outside of the growth zone (cambium) as a shell for the wooden body.	EN ISO 16559:2014
Wood-processing industry by-products	*Woody biomass* by-products originating from the wood processing as well as the pulp and paper industry.	FAO, UBET, 2004
	Cross-cut ends: short pieces of *woody biomass* which occur when the ends of logs or sawn timber are cross cutoff, with or without *bark*.	EN ISO 16559:2014
	Edgings: parts of *woody biomass* which occur when trimming sawn timber and which show a remainder of the original rounded surface of the tree, with or without *bark*.	EN ISO 16559:2014
	Slabs: parts of *woody biomass* created when cuts are made into the edges of logs and whereby one side shows the original rounded surface of the tree, either completely or partially, with or without *bark* (approximate length 200–800 cm).	EN ISO 16559:2014
	Sawdust: Fine particles created when sawing wood. Most of the material has a typical particle length of 1–5 mm.	FAO, UBET, 2004
	Wood shavings; cutter shavings: shavings from *woody biomass* created when planning wood.	FAO, UBET, 2004
	Black liquor: liquor obtained from wood during the process of pulp production, in which the energy content is mainly originating from the content of lignin removed from the wood in the pulping process. Black liquor also contains pulping chemicals. Black liquor is not a *solid biofuel*.	EN ISO 16559:2014

- other residues conifer,
- other residues nonconifer.
- Residues from industry-producing semifinished wood-based panels (veneer sheets, plywood, particle board, OSB, MDF, hardboard, and insulating board).
- Residues from further processing (construction, packaging, furniture, other).
- Residues from pulp and paper.

6.4.2.1 Saw Mill Residues

The basic methodological approach is to determine the amounts of by-products or residues based on the statistics on production volumes provided by FAOSTAT per country in combination with product recovery rates.

If data on production and consumption of sawn wood are not reliable, an alternative is to use the following formulae for calculation of P–Q (product quantity) using national-level data sources:

$$\mathrm{P} - \mathrm{Q} = \mathrm{AC} - \mathrm{I} + \mathrm{E}$$

where:

AC is apparent consumption, I is import, and E is export.

The amounts of sawdust and other residues from sawmills are estimated using:

$$\mathrm{SDQ} = \mathrm{PQ} \times \mathrm{SDP\ ratio}$$

$$\mathrm{ORQ} = \mathrm{PQ} \times \mathrm{ORP\ ratio}$$

where:

SDQ is standing for the sawdust quantity, PQ for the product quantity, SDP ratio for the sawdust to product ratio, and ORP ratio for the other residues to product ratios.

These ratios are determined using the recovery rate of the product from the roundwood input and the share of sawdust and other residues that are provided by UNECE/FAO (2010) for selected countries and by Saal (2010a,b) using:

$$\mathrm{RR}\% = \mathrm{PQ}/\mathrm{IQ}$$

$$\mathrm{SD}\% = \mathrm{SDQ}/\mathrm{IQ}$$

$$\mathrm{OR}\% = \mathrm{ORQ}/\mathrm{IQ}$$

where:

IQ stands for input quantity, RR for recovery rate, SD% for share of sawdust, and OR% for share of other residues, and the ratios result from:

$$\mathrm{SDP\ ratio} = \mathrm{SD}\%/\mathrm{RR}\%$$

$$\mathrm{ORP\ ratio} = \mathrm{OR}\%/\mathrm{RR}\%$$

The rates per country utilized and further details of the methodology to assess residues from sawmills are presented in Dees et al. (2017a).

6.4.2.2 *Residues From Semifinished Wood-Based Panels*

Residues from *semifinished wood-based panels* are determined using FAOSTAT production data on panel production per product category, residue shares from product input quantity and a factor that relates round wood input to product quantities. Merely for a few countries, country specific factors and shares are available, thus average values were used for all countries (derived from UNECE/FAO, 2010 and Saal, 2010b).

Using

$$\mathrm{ResQ} = \mathrm{SBP\%}\ \mathrm{PQ} \times \mathrm{IPF}$$

where:

ResQ is quantity of residues; PQ is product quantity, and IPF is roundwood input to product factor,

The factors used are given in Table 6.4. The conversion from solid cubic meters to tonnes dry matter assumes moisture content for the products of 6% for veneer and plywood and of 7% or all types of boards.

Further details of the methodology to assess residues from semifinished wood-based panels are presented in Dees et al. (2017a).

6.4.2.3 *Secondary Residues From Further Processing*

Residues from further processing are determined for four different sectors:

1. Construction
2. Packaging
3. Furniture
4. Other

For these sectors EUROSTAT provides production data for single products (according to the PRODCOM list) and structural business statistics.

Following the approach of Saal (2010a,b), a relation between the number of employees and the amount of consumed wood products per sector was determined using German data on these sectors (Mantau and Bilitevski, 2010). Their factors based on the NACE 1 classification system have been adjusted in S2Biom to the NACE 2 classification (Table 6.7).

The amount of residues is determined using:

$$\mathrm{SBPQ} = \mathrm{Res\%} \times \mathrm{NE} \times \mathrm{EF}$$

where:

Res% is residue shares per roundwood input, NE is number of employees per sector, and EF is expansion factor wood consumption/employee.

The factors used are presented in Table 6.8.

TABLE 6.7 Factors to Determine Secondary Residues From Semifinished Wood-Based Panels

Product Category	Residues Share From Product Input (%)	m^3 Roundwood Input/m^3 Product	Tonnes Dry Matter/ m^3 Average Shipping Weight
Particle board	3.94	1.51	0.6147
OSB	9.80	1.63	0.5701
MDF	9.61	1.68	0.6631
Hardboard	11.61	2.03	0.8565
Insulating board	4.57	0.83	0.2771
Veneer/ plywood	45.00	1.87	0.5601

TABLE 6.8 Business Subsectors Expansion Factors

Sector	Wood Consumption (m^3) per Employee (NACE 2)	Residues Shares (%) (Mantau and Bilitevski, 2010)
Construction	311.8	10.3
Furniture	79.4	18.4
Packaging	540.1	9.7
Other	117.5	13.0

6.4.2.4 Residues From Pulp and Paper Industry

Residues from pulp and paper industry include bark and black liquor. In a first step the roundwood (RW) input per pulp technology is estimated using:

$$\text{RWE input} = \text{P} \times \text{C}$$

where:

P is pulp in tonnes and C is conversion factor RW input per tonne.

Conversion factors used, published by UNECE/FAO (2010), are presented in Table 6.9.

According to Smoock (1992), approximately 40%–50% of the input raw material can be recovered as usable fiber. To estimate the amount of black liquor from the chemical pulping process, a factor of 0.5 is adequate to determine the amount of roundwood input that is included in the pulp. In addition, pulp contains chemicals to separate cellulose from lignin (Saal, 2010a).

For estimation of the amount of bark, the following formula is used:

$$B = RWE\text{-}I \times F\text{-share} \times B\text{-factor}$$

where:

B is bark volume, RWE-I is roundwood input (m^3), F-share is share of wood from forests, and B-factor is bark in relation to roundwood.

We used a factor for bark in relation to RWE of 10% considering an average over-bark/under-bark ratio of 0.88 (UNECE/FAO, 2010) and certain bark losses. Debarking in the forest is regarded negligible.

Bark residues result from debarking wood originating from roundwood. Since both residues from saw mills and roundwood are used by the pulp and paper industry, the share of wood from forests used in the pulp industry per countries is determining the amount of bark residues.

The shares per country have been determined using available industry statistics (Table 6.10).

TABLE 6.9 Conversion Factor Roundwood (input/tonne pulp)

Pulp Type	Conversion Factor (m^3 wood input/tonne pulp (dmt = air dried))
Mechanical	2.50
Semichemical	2.67
Chemical	4.49

TABLE 6.10 Share of Roundwood in the Pulp and Paper Industry

Country	Share (%)	Source
Austria	53.7	Value is based on data for Austria from Klima activ (2014)
Finland and Sweden	80.1	Value is based on data for Finland from Heinimö and Alakangas (2009)
Germany	66.0	66% is the average of annual values determined for Germany for 2010–14 using data from CEPI, Germany (2014) and CEPI, Germany (2012).
Other countries	75.0	Average share of CEPI countries and members determined from statistics by CEPI (2014).

6.4.3 Results

The base potential of secondary forestry residues in 2012 is dominated by residues from sawmill industry (over 40%), followed by pulp and paper residues with over 30%, whereas other wood-processing residues built the smallest but still considerable fraction with over 20%. More insight in the structure of the potentials by major category is provided in Fig. 6.8 for EU28.

The availability of secondary forestry residues is strongly depending on the wood industry structure per country. This gets visible in Fig. 6.9 that shows the secondary forestry residues potentials from saw mill residues, other wood-processing industries, and from pulp and paper industry for 2012–30. The highest amounts are available in Sweden, Finland, and Germany (together these countries provide nearly 50% of the potential of EU28) followed by Poland, France, and Austria (together these six countries provide about two-thirds of the potential of EU28), this in accordance with their well-established wood industry sectors.

The expected development per sector is based on the base scenario predictions of the study EFSOS II, except for the pulp and paper sector that is assumed to remain stable. In total there is in EU28 for 2020 an increase by 3.6% vs 2012 and for 2030 by 6.3% vs 2012. In the non-EU countries the increase is larger, toward 2020 on average it is 10.6% and toward 2030 20.0%.

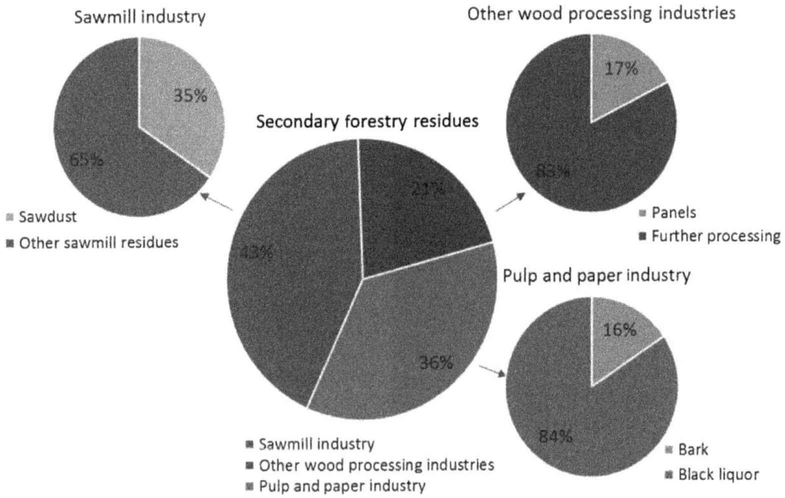

FIGURE 6.8 Supply from secondary forestry residues by major category—base potentials for all EU countries (2012).

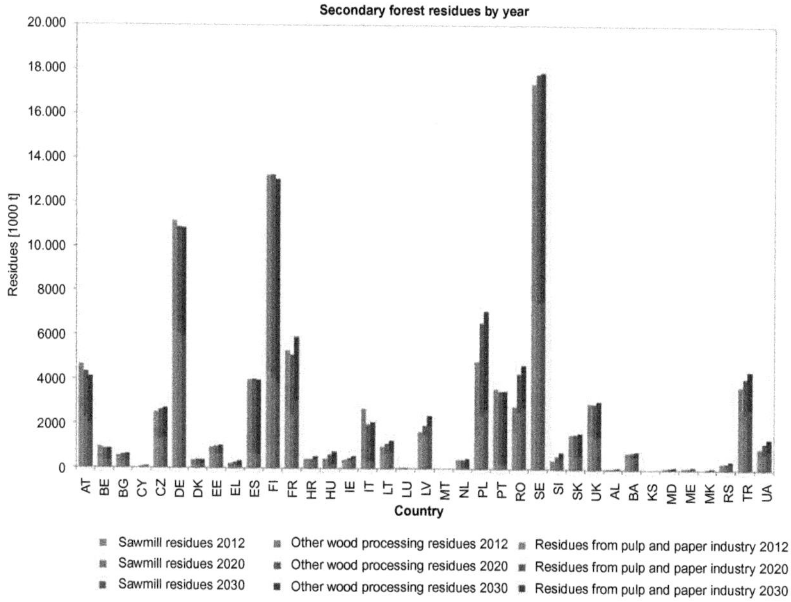

FIGURE 6.9 Secondary forest residue base potentials by country for 2012, 2020, and 2030.

6.5 REGIONAL FOCUS ON WESTERN BALKANS

The assessment of woody biomass potentials for the neighboring countries of the EU in the Western Balkans complemented the S2Biom results with additional analysis carried out in a recent FAO project with methods as described in the BEE project (Vis and Dees, 2011). The analysis of the total biomass potentials also included the analysis of potentials outside forests. To that effect, for countries that lacked research on the subject, the initial element for estimating potential of woody biomass outside forests was the other wooded land expressed in hectares and growing stock of other wooded land in cubic millimeter. Results in Serbia showed that the annual increment of trees outside forests was 1.9 million m^3 or 5.1% of the total growing stock of other wooded land. This amount was obtained by using the data from the national forest inventory and the results of the analysis of 4665 sample plots with radius of 250 m for the whole country (sample grid of 4×4 km^2). Furthermore, Corine land maps and satellite ortophotos coverage from Google Earth were used.

Using the methodology of Vis and Dees (2011) for the calculation of potential of biomass from forests and outside forest as well as numerous sources of the data for the needs of the Western Balkans Study, woody biomass potential was estimated both for each country individually and for the region as a whole. The total potential of woody biomass from forests and outside forest (without biomass from the conversion of coppices and from plantations) for energy purposes on the Western Balkans was 23.5 million m^3 (Table 6.11).

TABLE 6.11 Sustainable Potential of Woody Biomass for Energy Purposes From Forests and Outside Forest in the W-B

Country	Technical Potential—Stemwood for Energy[a]		Technical Potential of Logging Residue (m^3/year)	Technical Potential of Secondary Forestry Residue (m^3/year)	Wood Outside Forest Available for Wood Supply (m^3/year)	Total (m^3/year)
	Total (m^3/year)	Out of Which Thinning (m^3/year)				
Albania	659,233	35,000	57,197	57,138	167,314	940,882
Bosnia and Herzegovina	4,076,138	116,800	587,833	1,159,748	342,100	6,165,819
Croatia	3,494,631	128,000	659,112	1,011,104	75,000	5,239,847
FYROM	1,054,457	11,400	86,679	56,107	71,500	1,268,743
Kosovo	938,023	5,000	72,311	21,241	45,006	1,076,581
Montenegro	1,468,081	7,550	135,325	141,190	68,750	1,813,346
Serbia	4,684,593	110,960	483,176	440,684	1,441,427	7,049,881
Total W-B Region	16,375,156	414,710	2,081,633	2,887,212	2,211,097	23,555,099

[a] *Technical potential of stemwood for energy includes wood from natural and artificially established stands where thinnings are occasionally carried out. Amount of wood obtained from thinning varies for by countries and years. For example, 966,225 m^3 were produced from the thinning in natural and artificially established stands in Serbia in 2013. In Albania, the timber provided from forest thinning is estimated at around 35,000 m^3/annually in 2011.*

Source: 1, Forest resources and consumption of forest products in countries in transition. 1990–2020. UNECE/FAO. 1996; 2. Forestry development strategy of Republic of Srpska 2011–2021. Ministry of agriculture. forestry and water management of Republic of Srpska. Banja Luka. 2012; 3. Forests in the ECE Region. ECE/TIM/SP/37. United Nations. 2015; 4. SYNENERGY study. CRES. 2012; 5. Wood biomass sector in Kosovo. FAO. 2015; 6. WISDOM for Montenegro. FAO. 2014; 7. Ministry of agriculture, forestry and water economy of Republic of Macedonia: Forestry sector in Macedonia. Skopje. 2015; 7. JWEQ. UNECE. Geneva. 2014; 8. FAO forestry data base. 2015. 9. WISDOM Croatia. FAO. 2009;10. Calculations of Glavonjić B.

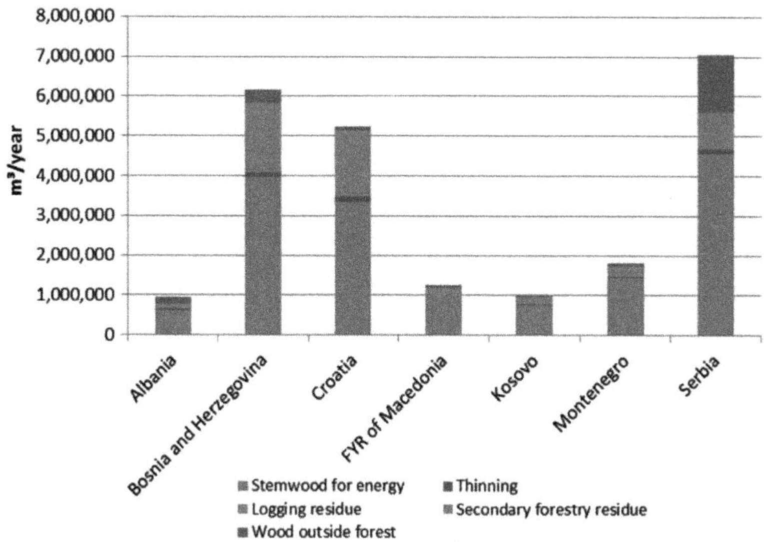

FIGURE 6.10 Sustainable potential of woody biomass for energy purposes from forests and outside forest in the Western Balkan region.

Forests are the most significant source of woody biomass in the Western Balkan region, contributing 16.2 million m^3 or 69% of the total potential. The remaining 31% is almost equally distributed among secondary forestry residue, logging residues, and wood outside forest available for wood supply (Fig. 6.10). For woody biomass potential from logging residues, the theoretical values were calculated first, and afterward the realizable potentials with consideration of technical, economical, and sustainable aspect were used.

Potentials outside forest are an important characteristic of woody biomass potential in the region. These potentials were estimated for certain countries as well as for the region as a whole for the first time. Serbia (20.4%) and Albania (17.4%) have the biggest share of potentials of woody biomass outside forest in their total forest biomass potentials, and Croatia (1.4%) and Montenegro (3.8%) have the lowest share. In other countries in the region, their share ranges about 5% while the average for the entire region is 9.4%. Having in mind the constant increase of demand for woody biomass on one hand and the possible use of these potentials on the other, it is conceivable that local residents will get increasingly interested in this potential in the near future. Use of these potentials for energy purposes in certain countries, like in Serbia, is highly expressed while in other countries it is not the case yet. There are various reasons for this, however the primary reason is that woody biomass is still locally available in forests in sufficient amounts.

Concerning secondary forestry residue, Croatia and Bosnia and Herzegovina have the most significant potentials, and Kosovo and Macedonia

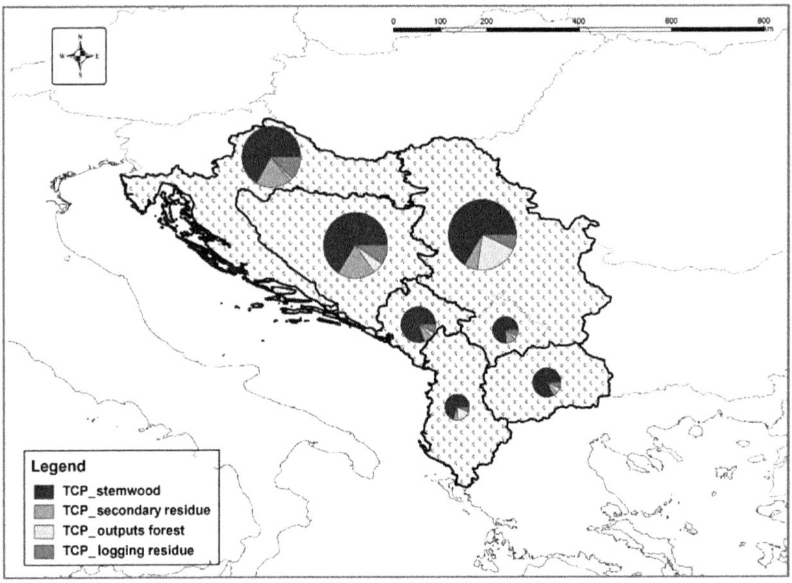

FIGURE 6.11 Share (%) of certain woody biomass categories in total potentials by countries.

have the least significant potentials. This is understandable as it corresponds to the amount of industrial roundwood processed annually in these countries. The situation is similar with the technical potential of logging residues. The leading country in the region is Croatia with 0.66 million m^3, followed by Bosnia and Herzegovina with 0.58 million m^3 and Serbia with 0.48 million m^3. Other countries are far behind these three countries (Fig. 6.11).

Serbia, Bosnia and Herzegovina, and Croatia have the biggest potentials of woody biomass and consequently the highest share. These three countries contribute 75.6% of the total woody biomass potential in the region, while the remaining four countries have significantly smaller share (Fig. 6.12). This analysis referred to technical potentials existing in forests and outside of forests, which is economically viable and sustainable taking into account established criteria for their sustainable use.

6.6 ACCESSIBILITY OF DATA

The data described in this chapter are accessible through the S2Biom toolset and database at http://biomass-tools.eu. The maps presented here represent only a small fraction of the overall available data. Biomass assessments can be carried out for different regions, selected tree species, specific biomass categories, as well as by type of potential, including additional user-defined potentials that were not described here for simplicity. All assumptions and a more detailed documentation of the methodology can be found in Dees et al. (2017b).

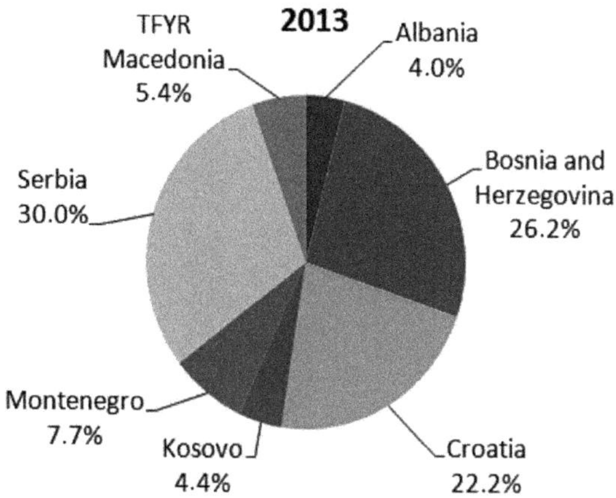

FIGURE 6.12 Share of individual countries in the total woody biomass potential of the Western Balkans region.

6.7 DISCUSSION AND CONCLUSIONS

The database described in this chapter is currently the most comprehensive source of information for forest biomass potentials and supply costs in Europe. The main strength compared to earlier assessments is the fact that the user can extract a large range of different data sets with user-defined assumptions and constraints to the potential availability of biomass.

Any forest biomass assessment relies on the accuracy and availability of recent resource information from national forest inventories. The collection of recent resource information is a considerable effort. Without regular updates, the database may get outdated after a few years. Hence, biomass potential assessments at European level would benefit enormously from improved access to detailed and latest national forest inventory results (i.e., public access to detailed and updated databases) for all European countries.

The comparison of different types of biomass potentials indicated that biomass availability grossly differs between a *high potential* assessment with minimum environmental constraints and a *user-defined biomass potential* that applies strict sustainability criteria with enhanced targets for biodiversity protection. There are currently many demands placed on forest ecosystem provisioning. Stakeholders have different views on these often conflicting demands and can reflect on these by choosing a set of constraints according to their own (or their customers) preferences.

A second major improvement of the present forest biomass assessment is the fact that supply costs were explicitly calculated using best available information on regional differences in relevant cost components. A supply–cost assessment at this level of detail for almost the whole of Europe has not been done before. The NUTS3-level cost estimates reveal huge differences between and within the countries. The figures can be utilized in large-scale biomass supply–cost analyses necessary for international political processes. The results may also serve country-level decision making in countries lacking data and capacity for more accurate assessments. However, due to the vast geographic area leading to varying conditions and data quality, several weaknesses in the cost assessment should be recognized:

- Only one supply chain per biomass category was determined.
- Part of the machine-level input data were assumed constant among the countries.
- Labor costs were based on Eurostat data with rather broad work categories.
- Only one productivity model per biomass category and work phase was applied.
- The accuracy of the intensity and average diameter data is unknown.
- Estimation of stem, crown, and stump volumes was simplified so that only one species for each of the four country categories was selected.
- Generally much of the input data is difficult to obtain at regional or even country level.

It was not possible to quantify the impacts of these uncertainties on the cost estimates, but they are important targets for future improvements to the presented methodology.

The secondary forest residues assessment relies on the accuracy of the production statistics at FAOSTAT. Partly these data are subject to substantial underestimations, depending on the methodology and approach applied by the countries providing the data. In contrast to other forest sector data sources, e.g., on the forest growing stock, there is no information available that qualifies the accuracy or level of uncertainty of the provided data. Both improvements of the accuracy of the data and a qualification of the underlying methodology in tier levels would increase understanding of the resource potentials substantially.

Further the underlying rates and factors need attention. Product recovery rates change with the technological development and need to be regularly updated. Moreover, they differ from country to country, but have not been available for all countries—for some categories they were reported only for very few countries.

An important lesson from the S2Biom analysis is the importance of regional difference between countries in Europe. The detailed assessment for the Western Balkan region indicated how different the potentials are within the region, but more importantly between this region and other parts of Europe. The comparison across the 37 investigated countries revealed that a few countries stand out in the amount of available biomass. Germany, Sweden, France, and Finland contribute the majority of forest biomass potentials in the whole region. In many other countries, the potentials are quite limited. This diversity will need to be reflected in national and regional policy strategies, e.g., in relation to the future utilization of forest biomass as a feedstock for a growing bioeconomy.

ACKNOWLEDGMENTS

This chapter is an output of the S2BIOM project that was funded through the European Union (FP7 Grant Agreement 608622). We would like to thank national forest inventory agencies for making available their data online or on request. Dragan Borota, Emil Cienciala, Lesley Halsall, Markus Höhl, Damjan Pantic, Mart-Jan Schelhaas, and Tetiana Zheliezna are acknowledged for their help in inventory data processing. The following experts kindly contributed to the forest cost data compilation: Dimitris Athanassiadis, Stelian Alexandru Borz, Anjo de Jong, Mehmet Eker, Lars Eliasson, Benjamin Engler, Jiri Gol, Nike Krajnc, Martin Kühmaier, Andi Lazdins, José Lousada, Martin MacPherson, Ivailo Markoff, Peeter Muiste, David Sanchez, Colin Saunders, Raffaele Spinelli, Aleksandar Stijović, Kjell Suadicani, Marijan Susnjar, and Tetiana Zheliezna.

REFERENCES

Ackerman, P., Belbo, Helmer, Eliasson, Lars, de Jong, Anjo, Lazdins, Andis, Lyons, John, 2014. The COST model for calculation of forest operations costs. Int. J. Forest Eng. 25 (1), 75–81. Available from: http://dx.doi.org/10.1080/14942119.2014.903711.

Anttila, P., Asikainen, A., Laitila, J., Broto, M., Campanero, I., Lizarralde, I., et al., 2011. Potential and supply costs of wood chips from forests in Soria, Spain. Forest Syst. 20 (2), 245–254.

Asikainen, A., Liiri, H., Peltola, S., Karjalainen, T., Laitila, J., 2008. Forest Energy Potential in Europe (EU27). Working Papers of the Finnish Forest Research Institute 69. Finnish Forest Research Institute, Helsinki, p. 33.

Brandini, P., Tabacchi, G., 1996. Biomass and volume equations for holm oak and strawberry-tree in coppice stands of Southern Sardinia. ISAFA Communicazioni di Ricerca 96 (1), 59–69.

Brus, D., Hengeveld, G., Walvoort, D., Goedhart, P., Heidema, A., Nabuurs, G., et al., 2012. Statistical mapping of tree species over Europe. Eur. J. Forest Res. 131, 145–157.

CEPI, 2014. Historic statistics 1991–2014.

CEPI, Germany, 2012. Consumption of pulpwood in the pulp and paper industry.

CEPI, Germany, 2014. Consumption of pulpwood in the pulp and paper industry.

Černý, M., 1990. Biomass of *Picea abies* (L.) karts in Midwestern Bohemia. Scand. J. Forest Res. 5, 83–95.

Corona, P., Ferrara, A., 1987. Dendrometrical investigations on *Pinus silvestris* in Trentino-Alto Adige. Monti e Boschi 38 (6), 51–54.

Dees, M. et al., 2015. Regional data collection in support of the data base on sustainable ligno-cellulosic biomass cost–supply. S2Biom project deliverable D1.1 v. 0.1. Manuscript.

Dees, M., Elbersen, B., Fitzgerald, J., Vis, M., Anttila, P., Forsell, N., et al., 2017a. A spatial data base on sustainable biomass cost–supply of lignocellulosic biomass in Europe—methods & data sources. Project Report. S2Biom—a project funded under the European Union 7th Framework Programme for Research. Grant Agreement no. 608622. Chair of Remote Sensing and Landscape Information Systems. Institute of Forest Sciences, University of Freiburg, Freiburg, Germany, p. 174.

Dees, M., Elbersen, B., Fitzgerald, J., Vis, M., Anttila, P., Forsell, N., et al., 2017b. Atlas with regional cost supply biomass potentials for EU 28, Western Balkan Countries, Moldavia, Turkey and Ukraine. Project Report. S2BIOM – a project funded under the European Union 7th Framework Programme for Research. Grant Agreement no 608622. Chair of Remote Sensing and Landscape Information Systems, Institute of Forest. Sciences, University of Freiburg, Germany, p.105.

DG Energy, 2015. Oil bulletin, Directorate-General for Energy, European Commission. Available from: <https://ec.europa.eu/energy/en/statistics/weekly-oil-bulletin>.

Díaz-Yáñez, O., Mola-Yudego, B., Anttila, P., Röser, D., Asikainen, A., 2013. Forest chips for energy in Europe: current procurement methods and potentials. Renew. Sustain. Energ. Rev. 21, 562–571. Available from: http://dx.doi.org/10.1016/j.rser.2012.12.016.

Eliasson, L., Granlund, P., von Hofsten, H., Björheden, R., 2012. Studie av en lastbils monterad kross—CBI 5800—study of a truck-mounted CBI 5800 grinder. Arbetsrapport från Skogforsk 775, 16.

Eriksson, G., Bergström, D., Nordfjell, T., 2013. The state of the art in woody biomass comminution and sorting in Northern Europe. Int. J. Forest Eng. 24 (3), 194–215.

European Central Bank, 2013. Press release. Euro area MFI interest rate statistics, 10 pp. Available from: <http://www.ecb.europa.eu>.

EUROSTAT, 2015a. Purchasing power parities (PPPs), price level indices and real expenditures for ESA2010 aggregates [prc_ppp_ind] Internet site. Available from: <http://ec.europa.eu/eurostat/cache/metadata/en/prc_ppp_esms.htm> (accessed 04.11.15).

EUROSTAT, 2015b. Labour cost, wages and salaries (including apprentices)—NACE Rev. 2 [lc_ncostot_r2]. Internet site. Available from: <http://ec.europa.eu/eurostat/cache/metadata/en/lcs_esms.htm> (accessed 15.04.15).

Fiedler, F., 1986. Die Dendromasse eines hiebsreifen Fichtenbestandes.. Beiträge für die Forstwirtschaft 20 (4), 171–180.

Forest Europe, 2015. State of Europe's Forests 2015. Madrid, p. 312.

Forest Europe, UNECE, FAO, 2011. State of Europe's Forests 2011. Status and Trends in Sustainable Forest Management in Europe, Oslo, p. 337.

Heinimö and Alakangas, 2011. Market of biomass fuels in Finland 2009.

Holzleitner, F., Kanzian, C., Höller, N., 2013. Monitoring the chipping and transportation of wood fuels with a fleet management system. Silva Fennica 47 (1), (article id 899, 11 pp).

Ilavský, J., Laitila, J., Tahvanainen, T., Tuček, J., Koreň, M., Pápaj, V., et al. 2007. Factors influencing availability of biomass resources and efficiency of its procurement for energy generation. A regional study for the Banská Bystrica Region, Slovakia. Working Papers of the Finnish Forest Research Institute 59. 57 pp. Available from: <http://www.metla.fi/julkaisut/workingpapers/2007/mwp059.htm>.

Junginger, M., Faaij, A., Björheden, R., Turkenburg, W.C., 2005. Technological learning and cost reductions in wood fuel supply chains in Sweden. Biomass Bioenergy 29 (6), 399–418.

Kindermann, G.E., Schörghuber, S., Linkosalo, T., Sanchez, A., Rammer, W., Seidl, R., et al., 2013. Potential stocks and increments of woody biomass in the European Union under different management and climate scenarios. Carbon Balance and Management 8.

Klima activ, 2014. Wood flows in Austria. <http://www.klimaaktiv.at/english/renewableenergy/woodflowsaustria.html>.

Kuitto P.-J., Keskinen S., Lindroos J., Oijala T., Rajamäki J., Räsänen T., et al., 1994. Puutavaran koneellinen hakkuu ja metsäkuljetus. [Mechanized felling and forest transport of timber.] Metsätehon tiedotus 410, 38 pp. (In Finnish).

Laitila and Väätäinen, 2011. Kokopuun ja rangan autokuljetus ja haketustuottavuus. [Truck transportation and chipping productivity of whole trees and delimbed stems.] Metsätieteen aikakauskirja 2/2011 (In Finnish).

Laitila J. (2010). Kantojen korjuun tuottavuus. [Productivity of stump harvesting.] Metlan työraportteja/Working Papers of the Finnish Forest Research Institute 150, 29 pp. (In Finnish). <http://www.metla.fi/julkaisut/workingpapers/2010/mwp150.htm> (In Finnish).

Laitila, J., 2006. Cost and sensitive analysis tools for forest energy procurement chains. Forestr. Stud. 45, 5–10.

Laitila, J., Ranta, T., Asikainen, A., 2008. Productivity of stump harvesting for fuel. Int. J. Forest Eng. 19 (2), 37–46.

Laitila, J., Ranta, T., Asikainen, A., Jäppinen, E., Korpinen, O.-J., 2015. The cost competitiveness of conifer stumps in the procurement of forest chips for fuel in Southern and Northern Finland. Silva Fennica 49 (2), (article id 1280, 23 pp.).

Laitila, J., Asikainen, A., Pasanen, K., 2012. Hankinnan teknologia, logistiikka ja hiilidioksidipäästöt. [Supply technology, logistics and carbon dioxide emissions.]. In: Asikainen, A., Ilvesniemi, H., Sievänen, R., Vapaavuori, E., Muhonen, T. (Eds.), Bioenergia, ilmastonmuutos ja Suomen metsät. [Bioenergy, Climate Change and the Finnish Forests.], 240. Metlan työraportteja/Working Papers of the Finnish Forest Research Institute, pp. 171–184. (In Finnish).

Laitila, J., Leinonen, A., Flyktman, M., Virkkunen, M. & Asikainen, A., 2010. Metsähakkeen hankinta- ja toimituslogistiikan haasteet ja kehittämistarpeet. [Challenges and development needs of supply logistics of forest chips.] VTT Tiedotteita 2564, 143 pp. (In Finnish).

Mantau, U. and BilitewskiI, 2010. Stoffstrom-Modell-Holz:Rohstoffströme und CO_2-Speicherung in der Holzverwendung.Celle:Forschungsbericht für das Kuratorium für Forschung und Technik des Verbandes der Deutschen Papierfabriken e.V. (VDP), 77 pp.

Meliadis, I., Zagkas, T., Tsitsoni, T., 2010. Greece. In: Gschwantner, T., Tomppo, E., Lawrence, M., McRoberts, R.E. (Eds.), National Forest Inventories. Pathways for Common Reporting. Springer, Heidelberg, pp. 423–437.

Menguzzato, G., Tabacchi, G., 1988. Modelli di previsione del peso fresco, della biomassa e del volume per pino insigne ed eucalitti nell'Azienda Massanova (Salerno). Ann. Ist. Sper. Selvicoltura 19, 323–354. (in Italian).

Mola-Yudego, B., Arevalo, J., Díaz-Yáñez, O., Dimitriou, I., Freshwater, E., Haapala, A., et al., 2017. Reviewing wood biomass potentials for energy in Europe: the role of forests and fast growing plantations. Biofuels, 1–10.

Nurminen, T., Korpunen, H., Uusitalo, J., 2009. Applying the activity-based costing to cut-to--length timber harvesting and trucking. Silva Fennica 43 (5), (article id 177). doi:10.14214/sf.177.

Nuutinen, Y., Laitila, J., Rytkönen, E., 2014. Grinding of stumps, logging residues and small diameter wood using a CBI 5800 grinder with a truck as a base machine. Baltic Forestr. 20 (1), 176–188.

Pellinen, P., 1986. Biomasseuntersuchungen im Kalkbuchenwald. Dissertation an der Univ, Göttingen, Germany, 145 pp.

Pretzsch, H., 2000. Die Regeln von Reineke, Yoda und das Gesetz der räumlichen Allometrie. Allgemeine Forst- und Jagdzeitung 171, 205–210.

Ranta, T., 2002. Logging residues from regeneration fellings for biofuel production—a GIS-based availability and supply cost analysis. Acta Universitatis 128. Lappeenranta University of Technology, 180 pp.

Ranta, T., 2005. Logging residues from regeneration fellings for biofuel production—a GIS-based availability analysis in Finland. Biomass Bioenergy 28, 171–182.

Repola, J., 2008. Biomass equations for Birch in Finland. Silva Fennica 42 (4), 605–624.

Repola, J., 2009. Biomass equations for Scots Pine and Norway Spruce in Finland. Silva Fennica 43, 4.

Rottensteiner, C., Tsioras, P., Neumayer, H., Stampfer, K., 2013. Vibration and noise assessment of tractor-trailer and truck-mounted chippers. Silva Fennica 47 (5), (article id 984, 14 pp).

Routa, J., Asikainen, A., Björheden, R., Laitila, J., Röser, D., 2013. Forest energy procurement—state of the art in Finland and Sweden. WIREs Energ. Environ. 2 (6), 602–613.

Saal, U., 2010a. Industrial wood residues. In: Mantau, U., Saal, U., Prins, K., Steierer, F., Lindner, M., Verkerk, H., et al. (Eds.), Euwood—Real Potential for Changes in Growth and Use of EU Forests. Final Report. Centre of Wood Science, University of Hamburg, Hamburg, Germany, pp. 97–107.

Saal, U., 2010b. Industrial wood residues. In: Mantau, U., Saal, U., Prins, K., Steierer, F., Lindner, M., Verkerk, H., et al. (Eds.), Euwood—Real Potential for Changes in Growth and Use of EU Forests. Methodology Report. Centre of Wood Science, University of Hamburg, Hamburg, Germany, pp. 124–145.

Sallnäs, O., 1990. A matrix model of the Swedish forest. Studia Forestalia Suecica 183, 23.

Schelhaas, M.-J., Eggers, J., Lindner, M., Nabuurs, G.-J., Päivinen, R., Schuck, A., et al., 2007. Model Documentation for the European Forest Information Scenario Model (EFISCEN 3.1.3). Alterra Report 1559 and EFI Technical Report 26. Alterra and European Forest Institute, Wageningen and Joensuu, p. 118.

Smook, G.A., 1992: Handbook for Pulp & Paper Technologists, second ed.

Teobaldelli, M., Somogyi, Z., Migliavacca, M., Usoltsev, V.A., 2009. Generalized functions of biomass expansion factors for conifers and broadleaved by stand age, growing stock and site index. Forest Ecol. Manage. 257, 1004–1013.

UNECE/FAO, 2010. Forest product conversion factors for the UNECE region. Geneva Timber and Forest Discussion Paper 49 Geneva, Switzerland, UNECE/FAO Timber Section. <http://www.unece.org/fileadmin/DAM/timber/publications/DP-49.pdf>.

UNECE and FAO (Eds.), 2011. The European Forest Sector Outlook Study II. 2010–2030, Geneva, Switzerland.

Verkerk, H., Immonen, V., Hengeveld, G., Kiljunen, J., Lindner, M., Luyssaert, S., et al., 2016b. Description and manual of open structure of models. Trees4Future Deliverable D10.4.

Verkerk, P.J., 2015. Assessing impacts of intensified biomass removal and biodiversity protection on European forests, Dissertationes Forestales 197, Joensuu.

Verkerk, P.J., Anttila, P., Eggers, J., Lindner, M., Asikainen, A., 2011. The realisable potential supply of woody biomass from forests in the European Union. Forest Ecol. Manage. 261, 2007–2015.

Verkerk, P.J., Schelhaas, M.J., Immonen, V., Hengeveld, G., Kiljunen, J., Lindner, M., et al., 2016a. Manual for the European Forest Information Scenario Model (EFISCEN 4.1). EFI Technical Report 99. European Forest Institute, Joensuu, p. 49.

Vis, M., Dees, M. (Eds.), 2011. Biomass Resource Assessment Handbook: Harmonisation of Biomass Resource Assessments, Best Practices and Methods Handbook. VDM Verlag Dr. Müller, Saarbrücken, 256 pp.

Wolfsmayr, U.J., Rauch, P., 2014. The primary forest fuel supply chain: a literature review. Biomass Bioenergy 60, 203–221.

FURTHER READING

Alakangas, E. & M. Virkkunen, 2007. Biomass fuel supply chains for solid biofuels. EUBIONET2 project (EIE/04/065/S07.38628), 32 pp.

Biomass Energy Europe (BEE), 2010. Harmonization of biomass resource assessment, Volume I. Best Practices and Methods Handbook. <http://www.eu-bee.info>.

European Commission, 2009. Natura 2000 Sites, Version January 2009. European Commission—DG Environment, Brussels.

Finnish Forest Research Institute, Helsinki.

Houšková, B., 2008. Natural Susceptibility of Soils to Compaction. European Commission, Institute of Environment and Sustainability, Land Management and Natural Hazards Unit, Ispra.

IPCC Guidelines for National Greenhouse Gas Inventories, 2006. Available from <http://www.ipcc-nggip.iges.or.jp/public/2006gl/>.

Jurevics, A., 2010. A Comparison of Harvesting Residue Yield and Recovery Rates for Energy Policy Development. Swedish University of Agricultural Sciences.

Levers, C., Verkerk, P.J., Müller, D., Verburg, P.H., Butsic, V., Leitão, P.J., et al., 2014. Drivers of forest harvesting intensity patterns in Europe. Forest Ecol. Manage. 315, 160–172.

Mantau, U., Saal, U., Prins, K., Steierer, F., Lindner, M., Verkerk, H. (Eds.), 2010. Euwood—Real Potential for Changes in Growth and Use of EU Forests, Methodology Report. Centre of Wood Science, University of Hamburg, Hamburg, Germany.

Panagos, P., 2006. The European soil database. GEO: Connexion 5, 32–33.

Riley, S.J., Degloria, S.D., Elliott, R., 1999. A terrain ruggedness index that quantifies topographic heterogeneity. Intermountain J. Sci. 5, 23–27.

WISDOM, 2015 Serbia, FAO.

World Bank, 2015. Sector study on biomass-based heating in the Western Balkans.

	Harvester	Excavator	Forwarder	Chipper	Grinder
Machine utilization (%)	80	88	85	65	65
Fuel consumption (l/h)	11.0	18.0	9.5	45.0	68.6
Insurance (€)	3750	1800	2500	5400	7300
Oil and lubricant cost (% of fuel cost)	15	5	5	15	15
Maintenance and repair cost (% of fuel cost)	51	34	34	41	41

		Machine Price After Index (€)				
	Machine Price Level Indices (2012)	Harvester	Excavator	Forwarder	Chipper	Grinder
Belgium	103	341210	163781	227473	491343	664223
Bulgaria	92.4	306095	146926	204064	440777	595866
Czech Republic	98.4	325972	156466	217314	469399	634558
Denmark	122.6	406140	194947	270760	584841	790618
Germany	97.7	323653	155353	215769	466060	630044
Estonia	96.2	318684	152968	212456	458905	620371
Ireland	105	347836	166961	231890	500883	677120
Greece	112.1	371356	178251	247571	534753	722906
Spain	98.2	325309	156148	216873	468445	633269
France	98.9	327628	157261	218419	471784	637783
Italy	100	331272	159011	220848	477032	644876
Cyprus	105.3	348830	167438	232553	502314	679055
Latvia	99.8	330610	158693	220406	476078	643587
Lithuania	93.4	309408	148516	206272	445548	602314
Luxembourg	100.8	333922	160283	222615	480848	650035

Hungary	89.4	296157	142155	197438	426466	576519
Malta	110	364399	174912	242933	524735	709364
Netherlands	102.3	338891	162668	225928	488004	659708
Austria	102.4	339223	162827	226148	488481	660353
Poland	94.8	314046	150742	209364	452226	611343
Portugal	106.1	351480	168710	234320	506131	684214
Romania	96.5	319678	153445	213118	460336	622306
Slovenia	96.9	321003	154081	214002	462244	624885
Slovakia	105	347836	166961	231890	500883	677120
Finland	113.2	375000	180000	250000	540000	730000
Sweden	113	374337	179682	249558	539046	728710
United Kingdom	98.8	327297	157102	218198	471307	637138
Croatia	96.2	318684	152968	212456	458905	620371
Albania	95.7	317027	152173	211352	456519	617147
Bosnia and Herzegovina	100.1	331603	159170	221069	477509	645521
Macedonia	93.8	310733	149152	207155	447456	604894
Montenegro	97.6	323322	155194	215548	465583	629399

		Machine Price After Index (€)				
	Machine Price Level Indices (2012)	Harvester	Excavator	Forwarder	Chipper	Grinder
Serbia	90.3	299139	143587	199426	430760	582323
Kosovo	90.3	299139	143587	199426	430760	582323
Ukraine	96.5	319678	153445	213118	460336	622306
Turkey	98	324647	155830	216431	467491	631979
Moldova	96.5	319678	153445	213118	460336	622306

	Labor Cost (€/h)						
	Fuel Cost (€/l)	Felling	Stump Lifting	Forwarding	Chipping	Grinding	Interest Rate (%)
Belgium	1.16	42.92	42.92	35.75	42.92	42.92	3.03
Bulgaria	1.02	3.29	3.29	3.54	3.29	3.29	3.03
Czech Republic	1.16	9.95	9.95	9.38	9.95	9.95	3.03
Denmark	1.12	40.71	40.71	37.71	40.71	40.71	3.03
Germany	1.21	35.22	35.22	25.49	35.22	35.22	3.03
Estonia	1.10	8.55	8.55	8.75	8.55	8.55	3.03
Ireland	1.19	31.75	31.75	28.92	31.75	31.75	3.03
Greece	1.18	16.31	16.31	19.12	16.31	16.31	3.03
Spain	1.11	23.03	23.03	21.45	23.03	23.03	3.03
France	1.12	36.38	36.38	32.18	36.38	36.38	3.03
Italy	1.35	27.20	27.20	24.73	27.20	27.20	3.03
Cyprus	1.13	15.32	15.32	19.08	15.32	15.32	3.03
Latvia	1.08	5.88	5.88	6.90	5.88	5.88	3.03
Lithuania	1.05	5.72	5.72	6.64	5.72	5.72	3.03
Luxembourg	1.07	30.73	30.73	30.76	30.73	30.73	3.03
Hungary	1.10	7.81	7.81	7.48	7.81	7.81	3.03
Malta	1.13	11.57	11.57	10.65	11.57	11.57	3.03
Netherlands	1.16	33.90	33.90	28.54	33.90	33.90	3.03
Austria	1.13	33.01	33.01	27.89	33.01	33.01	3.03
Poland	1.04	7.74	7.74	7.01	7.74	7.74	3.03
Portugal	1.12	10.99	10.99	15.02	10.99	10.99	3.03

		Labor Cost (€/h)					
	Fuel Cost (€/l)	Felling	Stump Lifting	Forwarding	Chipping	Grinding	Interest Rate (%)
Romania	1.00	4.17	4.17	4.53	4.17	4.17	3.03
Slovenia	1.09	15.13	15.13	15.12	15.13	15.13	3.03
Slovakia	1.15	9.18	9.18	7.94	9.18	9.18	3.03
Finland	1.20	35.03	35.03	29.69	35.03	35.03	3.03
Sweden	1.25	41.70	41.70	33.82	41.70	41.70	3.03
United Kingdom	1.45	21.73	21.73	21.22	21.73	21.73	3.03
Croatia	1.12	8.48	8.48	9.63	8.48	8.48	3.03
Albania	1.29	4.96	4.96	4.83	4.96	4.96	3.03
Bosnia and Herzegovina	1.09	4.96	4.96	4.83	4.96	4.96	3.03
Macedonia	0.83	4.96	4.96	4.83	4.96	4.96	3.03
Montenegro	1.07	7.31	7.31	5.68	7.31	7.31	3.03
Serbia	1.15	4.96	4.96	4.83	4.96	4.96	3.03
Kosovo	1.00	4.96	4.96	4.83	4.96	4.96	3.03
Ukraine	0.67	2.30	2.30	2.30	2.30	2.30	3.03
Turkey	1.19	5.68	5.68	5.91	5.68	5.68	3.03
Moldova	0.72	2.30	2.30	2.30	2.30	2.30	3.03

	Harvester	Excavator	Forwarder	Chipper	Grinder
Belgium	113	90	80	189	243
Bulgaria	58	40	38	115	163
Czech Republic	70	51	48	136	189
Denmark	118	90	87	195	251
Germany	102	81	67	177	231
Estonia	67	48	46	129	180
Ireland	101	78	73	175	230
Greece	84	61	63	155	211
Spain	86	65	62	153	204
France	103	81	75	175	227
Italy	95	75	68	173	231
Cyprus	79	58	61	146	200
Latvia	65	45	45	126	177
Lithuania	62	44	42	121	170
Luxembourg	96	74	73	165	216
Hungary	63	46	43	124	174
Malta	77	55	52	143	197
Netherlands	102	79	72	175	229
Austria	100	78	70	172	225
Poland	65	46	43	125	173
Portugal	74	53	56	140	193
Romania	60	42	40	118	166
Slovenia	75	56	54	139	190
Slovakia	72	52	48	138	192
Finland	108	83	76	185	242
Sweden	117	92	81	198	256
United Kingdom	89	70	65	169	230
Croatia	67	49	47	130	182
Albania	65	48	43	133	189
Bosnia and Herzegovina	64	45	42	125	177
Macedonia	58	39	38	109	152

	Harvester	Excavator	Forwarder	Chipper	Grinder
Montenegro	65	47	43	127	177
Serbia	61	44	41	123	174
Kosovo	59	41	39	115	162
Ukraine	54	33	34	98	137
Turkey	65	47	44	131	184
Moldova	54	34	35	100	141

Slope category (%)	0–10	10–20	20–30	30–40	40–50
Driving speed factor	1.00	1.11	1.35	1.79	2.78

Group	Country	Broadleaved	Conifers
CEU	Austria	*Fagus* spp.	*Picea abies*
CEU	Belgium	*Fagus* spp.	*Picea abies*
CEU	Germany	*Fagus* spp.	*Picea abies*
CEU	Denmark	*Fagus* spp.	*Picea abies*
CEU	France	*Fagus* spp.	*Picea abies*
CEU	Luxembourg	*Fagus* spp.	*Picea abies*
CEU	Netherlands	*Fagus* spp.	*Picea abies*
EEU	Czech Republic	*Fagus* spp.	*Picea abies*
EEU	Croatia	*Fagus* spp.	*Picea abies*
EEU	Hungary	*Fagus* spp.	*Picea abies*
EEU	Moldova	*Fagus* spp.	*Picea abies*
EEU	Poland	*Fagus* spp.	*Picea abies*
EEU	Romania	*Fagus* spp.	*Picea abies*
EEU	Slovenia	*Fagus* spp.	*Picea abies*
EEU	Slovakia	*Fagus* spp.	*Picea abies*
EEU	Ukraine	*Fagus* spp.	*Picea abies*
NEU	Estonia	*Betula* spp.	*Picea abies*
NEU	Finland	*Betula* spp.	*Picea abies*

Group	Country	Broadleaved	Conifers
NEU	Ireland	*Betula* spp.	*Picea abies*
NEU	Lithuania	*Betula* spp.	*Picea abies*
NEU	Latvia	*Betula* spp.	*Picea abies*
NEU	Sweden	*Betula* spp.	*Picea abies*
NEU	United Kingdom	*Betula* spp.	*Picea abies*
SEU	Albania	*Quercus ilex*	*Pinus sylvestris*/Pinus radiata
SEU	Bosnia and Herzegovina	*Quercus ilex*	*Pinus sylvestris/Pinus radiata*
SEU	Bulgaria	*Quercus ilex*	*Pinus sylvestris/Pinus radiata*
SEU	Cyprus	*Quercus ilex*	*Pinus sylvestris/Pinus radiata*
SEU	Greece	*Quercus ilex*	*Pinus sylvestris/Pinus radiata*
SEU	Spain	*Quercus ilex*	*Pinus sylvestris/Pinus radiata*
SEU	Italy	*Quercus ilex*	*Pinus sylvestris/Pinus radiata*
SEU	Montenegro	*Quercus ilex*	*Pinus sylvestris/Pinus radiata*
SEU	Macedonia	*Quercus ilex*	*Pinus sylvestris/Pinus radiata*
SEU	Malta	*Quercus ilex*	*Pinus sylvestris/Pinus radiata*
SEU	Portugal	*Quercus ilex*	*Pinus sylvestris/Pinus radiata*
SEU	Serbia	*Quercus ilex*	*Pinus sylvestris/Pinus radiata*
SEU	Turkey	*Quercus ilex*	*Pinus sylvestris/Pinus radiata*
SEU	Kosovo	*Quercus ilex*	*Pinus sylvestris/Pinus radiata*

Chapter 7

Assessing Potentials for Agricultural Residues

Calliope Panoutsou, Christoforos Perakis, Berien Elbersen, Tetiana Zheliezna and Igor Staritsky

7.1 INTRODUCTION

Agricultural residues can derive from both primary harvesting and pruning activities on field as well as by-products/residues from processing crops in the respective agro-industries. In this chapter, they are defined as primary and secondary agricultural residues, respectively.

Primary agricultural residues are produced on field following crop harvesting (field residues) and regular tree pruning activities (arboricultural residues).

Field crops are producing two types of field residues, i.e., fresh or green dry residues. Green field crop residues, such as sugarbeet, are left in the field in fresh, succulent condition. These residues have high moisture content and are usually rotting in the field while occasionally some of them are used for animal feeding. Dry field residues derive from field crops, cultivated in various European regions, and they may come from small grain cereals (wheat, barley, oat, rye, and rice), maize, oil crops (sunflower, rapeseed, etc.), etc. These residues are incorporated into the soil, burned in the field or collected and used for various purposes.

Arboricultural residues are the prunings of grapes and trees such as apple trees, olive oil trees, pear trees etc., as well as from final disposal of old trees or from removal of whole grape or tree plantations for cleaning fields to be given to other more productive crops.

Secondary agricultural residues are produced in the (industrial) processing of agricultural crops into products.

The aim of the chapter is to present the key challenges, the methodologies, and relevant indicators for assessing potentials for agricultural residues. Detailed information on the use of data sources and methods are also included.

Modeling and Optimization of Biomass Supply Chains.
DOI: http://dx.doi.org/10.1016/B978-0-12-812303-4.00007-0

7.2 PRIMARY AGRICULTURAL RESIDUES

7.2.1 Definitions

An overview of the primary agricultural residues considered in this chapter, their classification, and their definitions are provided in Table 7.1. This classification has been further refined in Europe during 2008–16 in BEE, Biomass Futures, Biomass Policies, and S2Biom research projects. This work improved classification and links of European biomass assessments to the main statistical data sources in Europe on which future assessments for technical biomass potentials will rely on and be updated from.

Information presented in Table 7.1 is further organized into three groups (category, residue type, and crop origin). Information is also provided on the area where this biomass can be available, e.g., field, farm, and/or plant, informing whether the residue refers to bulk loose biomass or whether it is

TABLE 7.1 Overview of Biomass Categories Included in Primary Agricultural Residues

Category	Residue Type	Crop Origin	Definition
Dry field	Straw/ stubbles	Cereals	Dried stalks of cereals (including rice), rapeseed, and sunflower which are separated from the grains during the harvest. Available on field.
		Oil crops: rapeseed/ sunflower	
		Rice	
		Grain maize (stover)	Stover consists of leaves, stalks, and bare cobs from grain maize plants left in the field after harvest.
Green field		Sugarbeet leaves and tops	Sugarbeet leaves and tops are the harvest residues separated from the main crop, during harvest. Available on field.
Dry arboricultural	Pruning/ cutting	Apple, pear, and apricot pruning	The prunings and cuttings of fruit trees, vineyards, olives, and nut trees are woody residues produced after cutting, mulching, and chipping activities. They are the result of normal pruning management needed to maintain the orchards and improve their productivity. Available on field.
		Cherries	
		Vineyards	
		Olives	
		Citrus	
		Nut trees	

of standardized form for which a developed market is already there and thus can be called a commodity.

This information is relevant as it improves understanding on how the residues can be further handled, pretreated, stored, and transported. Biomass available at centralized facilities (e.g., processing plants/mills) requires lower collection cost and biomass that is already a commodity that can be handled and stored in the framework of existing logistical arrangements.

7.2.1.1 Straw and Stubble Potentials From Arable Crops

Straw from small grain cereals and other arable crops is considered an agricultural residue with significant potential, primarily because cereals and oilseeds constitute major arable crops for Europe. However, concerns are expressed both for competition among current and emerging markets and for the sustainable removal rates.

If national and regional straw and stubble potentials are to be quantified, it is always advisable to take into account regional specific straw ratio factors which take account of local production and varieties used. These regional specific factors can be used to further improve the accuracy of the technical straw potentials.

7.2.1.2 Arboricultural Residues From Tree and Shrub Permanent Crops

Permanent crops deliver pruning residues throughout their lifetime and stump extractions when the plantation is at the end of its lifetime. Pruning activities form part of regular practice to enhance and maintain the production of the main crop.

In Europe, the most cultivated tree species delivering arboricultural residues are fruit trees (apple, pear, cherries, apricots, peach, etc.), vineyards, olives, citrus, berries, and nuts.

The following sections present the methodological context for estimating the technical potential for primary agricultural residues in Europe and provide detailed information for its current uses and sustainable removal rates.

7.2.2 Key Challenges for Primary Agricultural Residues

7.2.2.1 Currently Known and Emerging Uses

Current uses for primary agricultural residues include soil incorporation for maintenance of soil organic matter (SOM), animal bedding and feeding, as well as horticultural activities such as mushroom, flower bulbs, and strawberry production. (The EC Soil Framework Directive 52 defines SOM as "*the organic fraction of the soil, excluding undecayed plant and animal residues, their partial decomposition products, and the soil biomass.*" SOM consists of between 50% and 58% carbon (C) and hydrogen (H), oxygen (O), nitrogen (N), phosphorous (P), and sulfur (S). http://eur-lex.europa.eu/legal-content/ES/TXT/PDF/?uri = CELEX:52006PC0232&from = EN.) Emerging uses include renewable energy, second-generation biofuels, and bio-based

TABLE 7.2 Current and Emerging (in Bold) Uses for Primary Agricultural Residues

	Straw			Grain Maize	Sugarbeet Leaves and Tops	Prunings			
	Cereals	Oil Crops	Rice			Fruit Trees	Nut Trees	Olives	Vineyards
Incorporation	√	√	√	√	√				
Animal bedding	√	√	√						
Animal feed	√	√	√						
Mushroom production	√	√	√						
Frost protection	√	√	√						
Strawberry/lower bulbs production	√	√	√						
Paper and pulp	√	√	√						
Building materials	√	√	√			√	√	√	√
Energy, fuels	√	√	√	√		√	√	√	√

materials. Table 7.2 provides an overview of current uses for primary agricultural residues. Renewable energy from primary agricultural residues is referred here as emerging and includes the high-efficiency boilers and innovative CHP applications. It does not refer to old stoves and boilers with low efficiencies which use prunings. However, these traditional uses still exist and occur mostly in rural mountainous regions across Europe.

Straw: Straw is a field residue with many uses which provide an additional source of income to the farmer. This is particularly the case in countries and regions where there is a large concentration of livestock and/or other straw using horticultural activities. Furthermore, there are also countries where straw demand for nonagricultural uses, such as for electricity, heat, and advanced fuels (cellulosic ethanol), is already developing quickly. Examples of the latter include Denmark, Italy, and Spain.

Arboricultural residues: Like straw, arboricultural residues are also an important source of nutrients and carbon and should not be fully removed from the fields. Removal of part of the residues may still be possible without endangering the nutrient and carbon levels in the soil. However, knowledge on the levels of residue removal at which the carbon levels are maintained is limited. Overall it is however clear that leaving part of the (shredded) pruning residues will enhance the maintenance of soil organic carbon (SOC) levels. What these levels should be is difficult to determine given the wide diversity in permanent crop types, soil, and climate circumstances. The best is therefore to take local practices into account aimed amending the organic levels in the soils.

On the other hand, it should be mentioned that leaving pruning residues in the field can also be a source of infections for other crops, so applying appropriate pesticide treatment is advised to avoid such situations.

Another common practice for such residues is to burn them on open fires in the fields, but this practice is no longer permitted in all regions of Europe or it has recently become subject to stricter regulations.

7.2.2.2 Incorporating Residues Into the Soil and Providing a Source of Organic Nutrients

Incorporation of straw into the soil is a common practice to maintain SOM in the harvested fields. SOM is a particularly important soil quality and influences physical, chemical, and biological soil properties. These include the physical structure and stability, ease of cultivation, ease of root growth, water infiltration rate, erosion, nutrient uptake, and biodiversity. The practice of incorporation is considered one of the most effective soil management practices. It occurs following harvest and the degree to which straw is incorporated varies in different parts of the European Union (EU), and depends on soil conditions, farm type, and size. The option to incorporate is more prevalent in regions where there is a high level of knowledge and technical infrastructure. In some cases, e.g., small traditional farms in Spain, the lack of machinery or the cost to rent it can prevent ploughing-in the harvested straw. In Italy, good networks/cooperatives for machinery have been reported so straw ploughing-in is more common, even in smaller farms. (IEEP, 2012. Mobilising cereal straw in the EU to feed advances biofuel production. Available from: <http://www.ieep.eu/assets/938/IEEP_Agricultural_residues_for_advanced_biofuels_May_2012.pdf>.)

Further factors which affect the level of incorporation within an annual rotation are the soil moisture level, nitrogen content, and temperature. Because of these limitations, it is common for farmers, particularly in southern and Eastern Europe to incorporate straw into the soil much less frequently and occasionally in only 2 out of 3 years. In most cases, especially in semiarid climates, primary agricultural residues are entirely lost during the decomposition process and, therefore, their contribution to the organic content of the soil is negligible.

Of course, there is a positive contribution to the mineral equilibrium of the soil, especially phosphorus and potassium. The latter is more important for northern EU fields which are more potassium deficient compared to southern EU fields. Finally, the nitrogen contribution is small because agricultural residues are poor in nitrogen content.

Straw: Straw has a significant benefit for the farmer as a fertilizer source. Incorporating straw returns potash (K_2O) and to a lesser extent phosphate (P_2O_5) to the soil, reducing the costs of buying inorganic sources of these key nutrients. In addition, it contributes to the reduction in environmental impacts, such as nutrient runoff and greenhouse gas (GHG) emissions associated with fertilizer manufacture. The value of incorporating straw as a fertilizer source will vary on a site-by-site basis depending upon the soil type and fertilizer prices.

A recent study for the EC analyzed the effect of different management systems on SOM. (Gobin et al., 2011. Soil organic matter management across the EU best practices, constraints and trade-offs. Available from: <http://ec.europa.eu/environment/soil/som_en.htm>.) The study considered straw extraction rates of 30%, 50%, and 100% compared to a business as usual (BAU) extraction rate of 10% and estimated the potential impact on humified organic carbon content. The results for cereal straw show that the humified organic carbon is, respectively, 7%, 21%, and 38% lower than in the BAU scenario. In the case of rapeseed, the reductions are 9%, 22%, and 47%, respectively.

These results are representing EU averages. However, regional level about humified organic carbon levels at different extraction rates can be presented at high variation.

Arboricultural residues: Incorporation of arboricultural residues into the soil is not a recommended practice as their hardwood nature will make the decomposition process quite lengthy, especially in the southeast European countries where their presence is rather high.

7.2.2.3 Burning of Primary Agricultural Residues

In certain European countries, it is a common practice for primary agricultural residues to be burned in the field. This practice has been banned in the clear majority of Member States under compulsory Good Agricultural and Environmental Condition (GAEC) standards as applied under the Common Agricultural Policy (CAP). (Only four Member States do not impose a ban on the burning of arable stubbles under cross-compliance GAEC standards—Cyprus, France, Ireland, and Slovenia.) Some Member States, including Denmark and the United Kingdom, also ban straw and stubble burning under national legislation. There is a perception among farmers that straw burning increases soil fertility. This opinion is without any scientific justification and several worldwide data indicate the opposite.

In a classical work carried out for a period of 35 continuous years in Oregon (United States) by Oveson (1966) in an area with approximately 400 mm of annual precipitation, like the precipitation prevailing to several southern EU regions, it was observed a continuous decline of SOM, soil nitrogen, and grain yields during this 35-year period in a rotation fallow-wheat system. He also reported that incorporating straw into the soil has the same effect on SOM as burning. It seems that straw is entirely lost during the decomposition process. It should be also stressed out that straw losses during the decomposition process are higher under semiarid climates with well-aerated soils as is the case of many cereal growing fields of southern EU. As is shown in Figs. 7.1 and 7.2, straw burning, either in autumn or spring, has no different effects on yields compared to straw utilization. From the same figure it is shown that yields are kept constant if nitrogen is applied, something that is the usual practice throughout Europe. The explanation is that during the burning process virtually all the nitrogen is lost.

In addition, it should be noted that straw burning in southern countries is one of the main causes for fire outbursts in nearby agricultural crops and/or forests. Finally, the burning process is uncontrolled and therefore increases particulate, NO_x and hydrocarbon emissions.

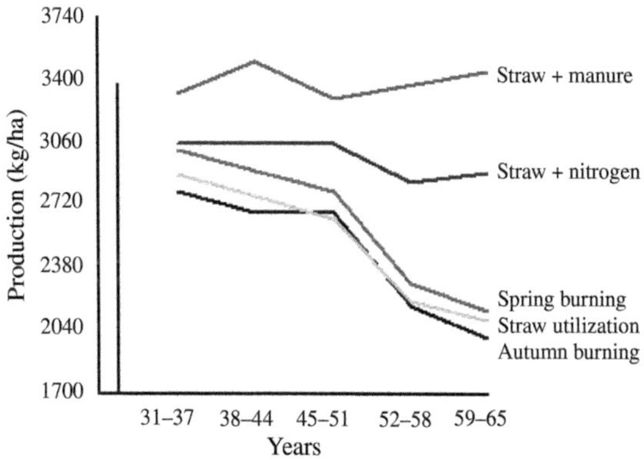

FIGURE 7.1 Winter wheat yields averaged over several straw residues treatments (Oveson, 1966).

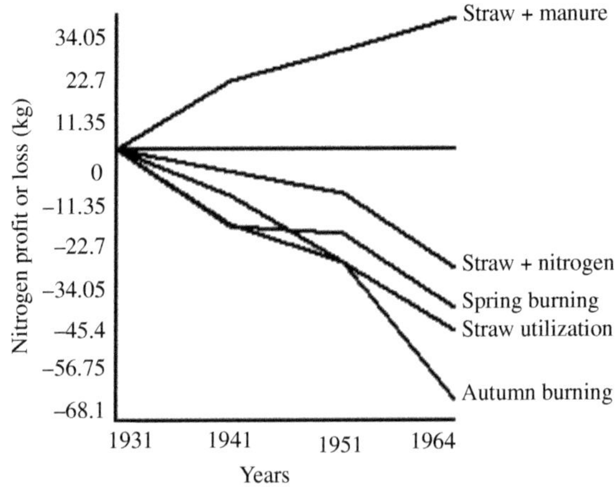

FIGURE 7.2 Nitrogen profit or losses estimated in the upper layer, up to 30 cm during a 34-year cultivating period averaged over several straw residue treatments (Oveson, 1966).

7.2.2.4 Effects on Soil Erosion

Erosion is a serious environmental hazard that reduces soil fertility as well as the productivity of agricultural land. It is reported that soil loss from erosion ranges from 11 to 20 t/ha/year (GAO, 1977; USDA, 1980c, 1981a), while soil is being formed under agricultural conditions at a rate of only 0.5–2 t/ha/year (Larson, 1981; Swanson and Harshbarger, 1964).

Soil erosion is more critical in the semiarid, sloping regions of southern European countries. The area of land with a high erosion risk in southern EU Member States is 22.9 million ha (about 10% of the rural land surface). The extent of this risk extends to one-third of land in Portugal, 20% in Greece, 10% in Italy, and 1% in France (Wieringa, 1995). A report by ICONA in Spain (1991) shows that some 44% of total land area is affected by erosion and 9 million ha (18%) currently loses more than 50 t/ha/year, which is considered the critical load for erosion. National average losses of 27 t/ha/year were reported, compared to soil formation of 2–12 t/ha/year.

Agricultural residues utilization after harvesting the crops in flat fields does not have any negative effect on soil erosion. Agricultural residues collection and utilization does not also have any negative effect if the field is planted with autumn sowing crops.

On the contrary, on the hilly, semiarid and sloping regions, especially of southern Europe, maintenance of agricultural residues in the field during the winter months may offer valuable protection against soil erosion if the field is planted with spring sown plants.

However, and contrary to the beneficial effects on soil erosion, even in this case farmers either burn or incorporate them into the soil in autumn. The reason being to allow for the incorporated residues to decompose during the winter and have their fields ready early in the spring. Therefore, even in this case, residues could be exploited for other purposes.

7.2.2.5 Impacts of Harvesting–Transportation Operations

Harvesting and transporting primary agricultural residues will increase the atmospheric emissions and the noise generated by vehicles. In addition, the extra traffic from such transport will increase traffic along specific routes, resulting in possible congestion and additional accidents. Therefore, to minimize those impacts, it is important to construct the processing plants at a relatively close distance from the raw material location (Bates et al., 1996).

7.2.2.6 Biomass Logistics

Straw: Residual biomass from small grain cereals and oil crops will be generally available in early autumn. Additionally, residues coming from early harvest hybrids of maize will also be available till the end of October. On the other and, late harvest hybrids of maize would be available in March–April, considering a period of 4–5 months for natural drying.

Farm ownership and structure and the large number of small-sized farms may pose some constrains regarding the collection of residues for central biomass installations. Small-scale or even individual biomass boilers may be more appropriate in some cases.

Arboricultural residues: Literature suggests that it may not be cost efficient to collect pruned material from plantations 0.5 ha or smaller, especially if the parcels are widely distributed. Plantations of this size are likely to be owned by farmers who do not rely entirely on farming income for their livelihood, cannot afford mechanical pruning contractor fees, and may not intensively manage their plantation. This may further limit the availability of pruning residues, since, as explained above, the proportion of small farms is quite big, especially in southeast parts of Europe.

Residues from pruning fruit trees consist of thin branches and leaves and thick branches (diameter > 3 cm). Thick branches are in most cases already used by farmers as fuelwood. However, this is not the case for smaller branches (including leaves), as special machinery is required for chipping and baling these residues, which of course cannot be afforded by most farmers. In this respect, a viable option could be machinery to be bought at a community level that would offer biomass collection services to interested farmers. Such cases can be examined at a local/community implementation level.

7.2.3 Methodology for Assessing Potentials From Primary Agricultural Residues

7.2.3.1 Data Sources

Data on area and yields of the main straw delivering crops can be derived from national and European-wide agricultural statistical sources (Table 7.3).

TABLE 7.3 Overview of Data Sources Providing Information on Area and Yields and Can Be Used as Input for Calculation of Potentials From Primary Agricultural Residues

Data Source	Spatial Coverage	Spatial Resolution	Description/Relevance
FADN	EU28	Regional (NUTS2/3)	Farm accountancy data
EUROSTAT/FSS	EU28 + Norway, Switzerland, Croatia	Regional (NUTS2/3)	Data on areas under cultivation per crop
Eurostat annual crop statistics	EU28	National and for some items regional (NUTS1/2)	Crop statistics are collected on areas under cultivation (expressed in 1000 ha), the quantity harvested (expressed in 1000 tonnes) and the yield (expressed in 100 kg/ha)
IACS/LPIS	EU28	Parcel size	Land use per parcel to be aggregated to any regional levels. Disclosure rules make access to these data difficult in some countries

Average crop yield data are available at national level and it is recommended to use annual averages across a 5- to 10-year period to capture the respective weather and climatic variations.

7.3 FARM ACCOUNTANCY DATA NETWORK

The Farm Accountancy Data Network (FADN) is an annual survey carried out by all the Member States of the EU. FADN data are collected every year from a sample of the agricultural holdings which are selected based on sampling plans established at the level of each region in a given country. The methodology applied aims to provide representative data along three dimensions: region, economic size, and type of farming.

FADN represents the only source of harmonized microeconomic (farm level) that derives from the same methodologies across all countries. Respective aggregated data can be found in the Standard Results database (http://ec.europa.eu/agriculture/rica/diffusion_en.cfm).

Currently, the annual FADN sample covers approximately 80,000 holdings. They represent a population of about 5 million farms in the Member States, which cover approximately 90% of the total utilized agricultural area (UAA) and account for about 90% of the total agricultural production of the Union.

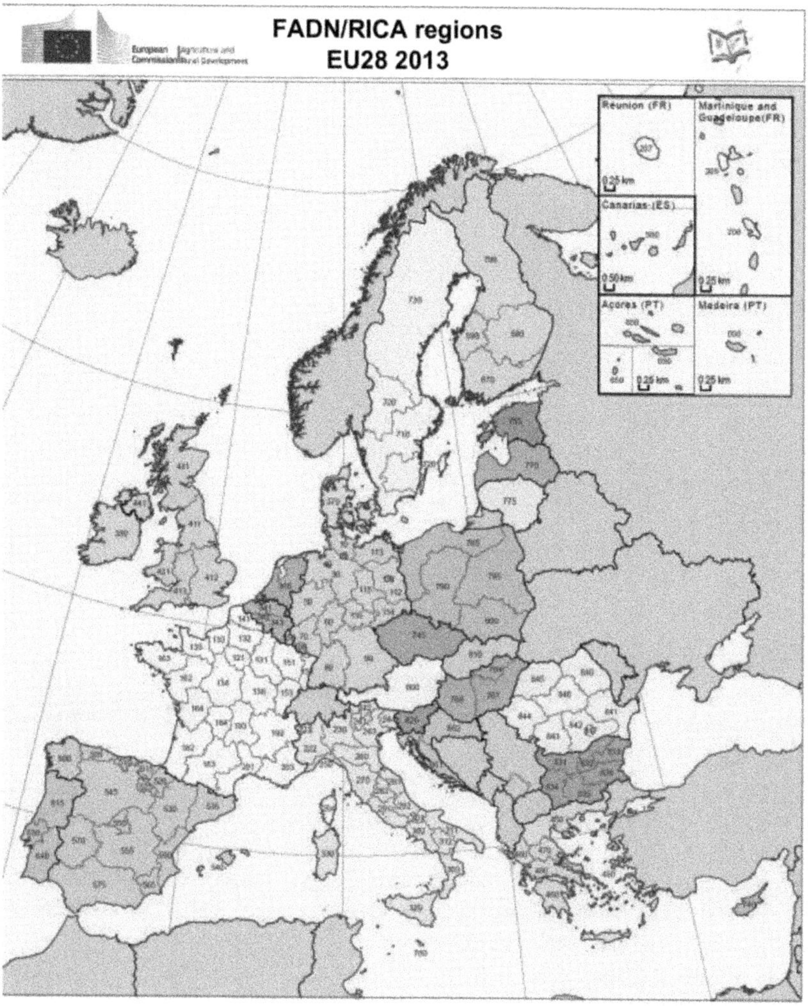

MAPS 7.1 FADN regions.

The stratification of the FADN sample is done using the information of the farm structure survey (FSS) on distribution of the total EU farm population over size and (sectoral) type. The survey however does not cover all the agricultural holdings in the Union but only those which due to their size could be considered commercial. This implies that often part-time farms and smaller extensive farms particularly occurring in Southern and CEEC are underrepresented.

The FADN data are available at individual farm level per FADN region. However, the accessibility to the individual farm data is restricted to FADN Liaison Agencies and permission to use the data needs to be obtained from DG-Agri. Data are only available at the FADN region level (Map 7.1).

Publication of the data in reports requires a minimal representation of the figure by more than 15 farms.

7.3.1 EUROSTAT Statistical Data Sources (FSS, Crop Survey, and SAPM)

The main EU-wide statistical source of information on agricultural land use is the FSS and the crop survey.

FSS provides EU-wide harmonized data on agricultural holdings in the EU on:

- number of agricultural holdings;
- land use and area (crops);
- main crops, area, yield, total production;
- farm labor force (including age, gender, and relationship to the holder);
- economic size of the holdings;
- type of activity;
- other gainful activity on the farm;
- system of farming;
- machinery;
- organic farming.

The frequency of data collection for FSS is every 2 years. This implies that the Member States are obliged to deliver the standard data every 2 years which can be based on a sample of farms but every 10 years a full scope survey is carried out in the form of an agricultural census. The most recent census took place in 2010 in all EU Member States.

The survey data can only be derived in aggregated format at different geographic levels (Member States, regions, and for basic surveys also district level). The data can also be arranged by size class, area status, legal status of the holding, objective zone, and farm type.

In addition to the FSS there is also *survey on agricultural production methods (SAPM)* which was carried out for the first time in 2010 to collect data at farm level on agri-environmental measures. EU Member States could choose whether to carry out the SAPM as a sample survey or as an exhaustive survey. Data were collected on tillage methods, soil conservation, landscape features, animal grazing, animal housing, manure application, manure storage and treatment facilities, and irrigation. Regarding irrigation, Member States were asked to provide estimation (possibly by means of models) of the volume of water used for irrigation on the agricultural holding.

The Member States collected information from individual agricultural holdings and, observing rules of confidentiality, data were transmitted to Eurostat. The results of the SAPM are linked at the level of individual agricultural holdings to the data obtained from the FSS in 2010. (The legal basis for the SAPM is Regulation 1166/2008 of November 19, 2008, on farm structure surveys and the survey on agricultural production methods, which repealed Council Regulation 571/88.)

The basic unit underlying the SAPM is the agricultural holding: a technical–economic unit, under single management, engaged in agricultural production. The SAPM covers all agricultural holdings with a UAA of at least 1 ha and those holdings with a UAA of less than 1 ha where their market production exceeds certain natural thresholds.

Crop statistics refer to annual data on area, production harvested and yield for cereals, and for other main field crops (mainly dried pulses, root crops, fodder, and industrial crops); humidity of the harvested crop (humidity content in %) and agricultural land use. For some products regional figures (NUTS1 or -2) are available too. The data refer to areas under cultivation (expressed in 1000 ha), the quantity harvested (expressed in 1000 tonnes), and the yield (expressed in 100 kg/ha). The information concerns more than 100 crop products.

The earliest data are available from 1955 for cereals and from the early 1960s for fruits and vegetables. However, most Member States have started to send in data in the 1970s and 1980s. The statistical system has progressively improved and enlarged. The current Regulation (EC) No. 543/2009 entered into force in January 2010. It simplified the data collection and reduced the number of crop subclasses. At present Eurostat receives and publishes harmonized statistical data from 28 Member States, which are broken down into:

- 17 categories and subcategories for cereals;
- 30 categories and subcategories for other main crops (mainly dried pulses, root crops, and industrial crops);
- 40 categories and subcategories for vegetables;
- 41 categories and subcategories for fruits;
- 18 categories and subcategories for UAA.

The main data sources are administrative records, surveys, and expert estimates. National Statistical Institutes or Ministries of Agriculture are responsible for the national data collection in accordance with EC Regulations. Eurostat is responsible for drawing the EU aggregations. For further information, see http://ec.europa.eu/eurostat/cache/metadata/en/apro_cpp_esms.htm#meta_update1418757870727.

Data for Western Balkans, Moldova, and Ukraine can be found on their national statistics and FAOSTAT.

7.3.2 Methodology

The methodology described in this section addresses the theoretical, technical, and sustainable potentials for primary agricultural residues in

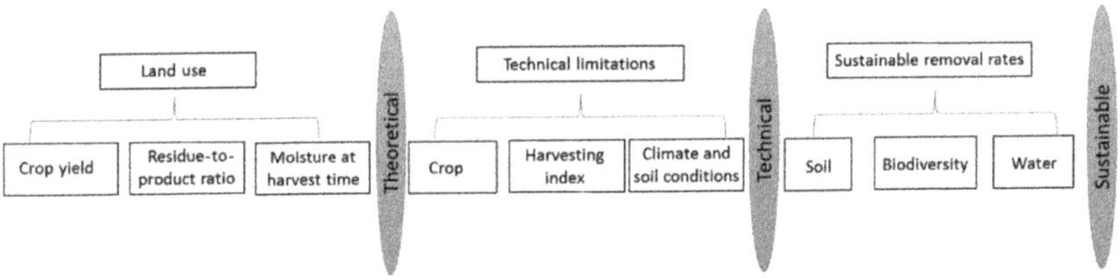

FIGURE 7.3 Methodology for assessing the potentials from primary agricultural residues.

Europe. Fig. 7.3 shows an overview of the methodological steps and the main assumptions in the calculation flows.

Theoretical potential: The total quantity of residues that can be produced annually from a specific crop. Theoretical potential is the quantity grown or disposed, constrained only by land, and crop growth–related factors.

For the theoretical potential, the calculation of the residue-to-yield factor (Table 7.4) is applied to the main product yield to estimate the aboveground biomass production per crop with the following formula:

$$\text{Theoretical residue yield}(i) = \text{Area }(i)^{*}\text{Yield (i)}^{*}\text{Residue ratio (i)} \\ ^{*}\text{Dry matter content }(i)$$

where:

Theoretical residue yield (i) = aboveground biomass of crop i
Area (i) = Crop area of crop i
Yield (i) = Yield level of the main product of crop i
Residue ratio (i) = Residue-to-main crop ratio for crop i
Dry matter content (i) = Dry matter content of crop i

It is worth noting here that estimating the theoretical residue yield per hectare depends strongly on local practices. However, to facilitate work in top-down biomass potential assessments, default values for the residue ratios can be found in relevant literature.

A set of representative values for residue–yield ratios and dry matter content (%) are presented in Tables 7.4 and 7.5.

Small grain cereals and oilseed crops are harvested during the summer period and therefore moisture content of crop residues is low (15% on average).

Maize is an exception, since it is harvested from second half of September to late November and therefore moisture content will depend on the cultivated hybrid as well as on weather conditions and whether they are suitable for on-field drying. Thus, moisture content of maize stover at harvest time will vary from less than 15% to over 50%. Nevertheless, if properly stored, moisture content of maize stover usually drops below 20% after a few months (4–5 months).

TABLE 7.4 Residue–Yield Ratios as Reported in Selected Studies for the Understudy Primary Agricultural Residues

	Straw			Grain Maize	Sugarbeet Leaves and Tops	Prunings			
Study (Country/ies)	Cereals	Oil Crops	Rice			Fruit Trees	Nut Trees	Olives	Vineyards
Panoutsou (EU)	1.0	1.8	0.8		0.70	0.35	0.35	1.3	0.5
Scarlat (EU)	0.8	2.0	0.8						
Synenergy (Western Balkans, UA, MD)	1.0	2.0		1.5		0.30	0.35	1.0	0.25
Geletukha (UA)	1.0	2.0	1.0		0.75	0.30	0.30		0.25
World Bank (Western Balkans)	1.0	1.5		1.0		0.35	0.20	0.70	0.40

TABLE 7.5 Dry Matter Content as Reported in Selected Studies for the Understudy Primary Agricultural Residues

	Straw			Grain Maize	Sugarbeet Leaves and Tops	Prunings			
Study (Country/ies)	Cereals	Oil Crops	Rice			Fruit Trees	Nut Trees	Olives	Vineyards
Panoutsou (EU)	75	60	75	60	50	80	80	80	80
Scarlat (EU)	85	60	75	70					
Synenergy (Western Balkans, UA, MD)	85	85		85	55	85	85	85	85
Geletukha (UA)	80	75	80	65	45	80	80	80	80
World Bank (Western Balkans)	85	85		80		80	80	80	80

Technical potential: This considers technical limitations related to the crop physiology, the harvest/collect index, and the prevailing climate and soil conditions in each region. There are several factors that limit the amount of primary field residues that can be recovered from the fields. These include the harvesting equipment, the crop type, growth pattern and variety, the harvest index (related to the growth and crop management practices), as well as losses from lodging (crops flattened by wind or rain). Thus, calculation of technical potential considers these factors and is lower than the theoretical one.

$$\text{Technical potential} = \text{Theoretical residue} - \text{yield}\,(i)^{*}\,\text{technical availability factor}$$

A set of representative values for technical availability is presented in Table 7.6.

Sustainable potential: The term "sustainable potential" is used to describe the amount of primary agricultural residues that can be removed from the land without adversely affecting the soil quality, causing negative impacts to biodiversity and water in each region. Consequently, the sustainable potential is lower than the technical one.

Regarding soil quality, appropriate accounting of residue removals from the field has to consider the following: removal of nutrients available in the harvested residues, impact on SOM, SOC, and effects on reduction or elimination of wind erosion and soil compaction protection cover. These should be considered to preserve soil fertility and productivity while in the same time allow for residue exploitation in the various markets.

$$\text{Sustainable potential} = \text{Technical residue} - \text{yield}\,(i)^{*}\,\text{sustainable removal rates.}$$

A set of representative values for sustainable removal rates is presented in Table 7.7.

7.4 SECONDARY AGRICULTURAL RESIDUES

Secondary agricultural residues derive from processing of primary agricultural crops into various commodities for the food, feed, and materials markets.

7.4.1 Definitions

There are several types of secondary agricultural processing residues, such as olive pomace and pits, cotton gin trash, almond shells, and peach pits.

An overview is presented in Table 7.8 along with detailed information is provided in the following section, including their currently known uses.

TABLE 7.6 Technical Availability Factors

	Straw			Grain Maize	Sugarbeet Leaves and Tops	Prunings			
Study (Country/ies)	Cereals	Oil Crops	Rice			Fruit Trees	Nut Trees	Olives	Vineyards
Panoutsou (EU)	50	70	50	45	45	85	85	65	70
Scarlat (EU)									
Synenergy (Western Balkans, UA, MD)	30	30		30	40	85	85	50	85
Geletukha (UA)	40	40	40	50	60				
World Bank (Western Balkans)	40	40		50		80	80	80	70

TABLE 7.7 Sustainable Removal Rates

	Straw			Grain Maize	Sugarbeet Leaves and Tops	Prunings			
Study (Country/ies)	Cereals	Oil Crops	Rice			Fruit Trees	Nut Trees	Olives	Vineyards
Panoutsou (EU)	35	45	35	50	55	85	85	85	75
Scarlat (EU)	40		40	45	50				
Synenergy (Western Balkans, UA, MD)									
Geletukha (UA)	30	40	30	40					
World Bank (Western Balkans)	40	40		50		80	80	80	70

TABLE 7.8 Overview of Residue Categories Included in Secondary Agricultural Residues

Category	Residue Type	Crop Origin	Definition
Secondary residues using agricultural products	Olive pits	Olive trees	Olive pits are separated from the rest of the olives when these are processed for making olive oil or when processed to tale olives
	Fruit pulp, kernels, husks, etc.	Fruit trees, vineyards, arable crops	Agro-industries process primary agricultural products into final products through pealing, crushing, drying, etc. and this results in many vegetal residues (e.g., soya and rape cake, and seed husks)

7.4.2 Key Challenges

Agricultural processing industries in Europe can be further grouped into the ones operating at industrial large scales and the ones that are more local, family operations. The first ones are quite common in north, west, and central part, while the second can be found in the south and eastern part of the continent. While exploitation of secondary agricultural residues is well developed in the first category, countries with high presence of small local and family agro-industrial enterprises are lagging behind. Knowledge and technology developments, and lack of awareness among primary producers and small scales make it difficult to improve the respective e-market development.

7.4.2.1 Currently Known and Emerging Uses

Table 7.9 presents currently known and emerging uses for secondary agricultural residues. These include fertilizers, animal feed, and comport for the first category; renewable energy and bio-based chemicals and materials for the second.

Renewable energy from secondary agricultural residues is referred here as emerging and includes the high-efficiency boilers and innovative CHP applications which produce processing heat and electricity within the respective agro-industries. It does not refer to old-type boilers with low efficiencies which use the secondary residues with low efficiencies. However, these traditional uses still exist and occur mostly in rural regions across Europe.

TABLE 7.9 Currently Known and Emerging (in Bold) Uses for Secondary Agricultural Residues

	Olive Pomace and Pits	Cotton Gin Residues	Cereal Bran	Rice Husks	Oil Crop Residues	Pressed Grapes	Fruit Juice Dregs	Sugarbeet Industries	Potato Industry Residues
Fertilizers						√			
Animal feed		√	√	√	√		√	√	
Compost		√	√		√	√	√	√	√
Energy	√								
Bio-based chemicals and materials	√	√	√			√	√	√	√

7.4.3 Proposed Resource Assessment Methodology

This section presents the methodology and indicators individually for each of the secondary residues included in this chapter.

7.4.3.1 *Olive Pomace and Pits*

Olive pits are a by-product from the olive oil industry. Calculation of the secondary residues from this source can be based on the total area of oil olives and the respective average yield per hectare of olive pits or the residue/main product ratio.

$$\text{Area} * \text{olive pit yield per hectare}$$

Suggestions for ratios:
*Italy (Blasi di et al., 1997): 0.3 tonne dry matter/ha/year of olive pits.
or

$$\text{Area} * \text{olives' yield per hectare} * \text{residue} - \text{to} - \text{main product ratio}$$

*Portugal (Diaz and Azavedo, 2004): 0.07 dry matter residue/main product ratio.

*Greece (Nikolaou et al., 2002): 0.21 wet mass residue/main product ratio (wet assuming 30% moisture).

7.4.4 Cotton Gin Residues

The process of cotton ginning produces a by-product composed of bur and stem fragments, immature cottonseed, lint, leaf fragments, and dirt, the so-called cotton gin waste or trash. These are residues from the cotton ginning factories.

Calculation of the secondary residues from this source can be done based on the total area (hectares) of cotton and the average per hectare yield of the main product, and then apply a residue/main product ratio.

$$\text{Area} * \text{per hectare yield of cotton} * \text{residue} - \text{to} - \text{main product ratio}$$

*Greece (Nikolaou et al., 2000): 0.1 wet mass residue/main product ratio (wet assuming 13% moisture).

7.4.5 Cereal Bran

These are residues from the flour mills. In wheat processing 20%–25% wheat is left unused (Kent et al., 1994). Wheat bran represents roughly 50% of these quantities and about 10%–19% of the kernel, depending on the variety and milling process (Ash, 1992; WMC, 2008; Prikhodko et al., 2009; Hassan et al., 2008). So the residue-to-yield factor used is 10% of cereals processed.

Calculation of the secondary residues from this source can be done based on the total area (hectares) of cereals used for flour production and the average per hectare yield of the main product, and then apply a residue/main product ratio.

$$\text{Area} * \text{per hectare yield of cereal} * \text{residue} - \text{to} - \text{main product ratio}$$

*Portugal (Diaz and Azavedo 2004): for rye 0.3 wet mass residue/main product ratio (wet assuming 12.5% moisture). For wheat flower the ratio is 0.19 (13% moisture) and for maize flower the ratio is 0.2 (18% moisture).

7.4.6 Rice Husk

These are residues from the rice mills. Rice husk is approximately 20% of the processed rice, with average moisture content of 10% (Nikolaou, 2002).

Calculation of the secondary residues from this source can be done based on the total area (hectares) of rice and the average per hectare yield of the main product, and then apply a residue/main product ratio.

$$\text{Area} * \text{per hectare yield of rice} * \text{residue} - \text{to} - \text{main product ratio}$$

*Greece (Nikolaou et al., 2002): 0.16 wet mass residue/main product ratio (wet assuming 10% moisture).

7.4.7 Oil Crop Residues (Soy, Sunflower, and Rape Seed)

These are residues from seed oil processing factories.

Calculation of the secondary residues from this source can be done based on the total area (hectares) of the crops and the average per hectare yield of the main product, and then apply a residue/main product ratio.

$$\text{Area} * \text{per hectare yield of oil seeds} * \text{residue-to-main product ratio}$$

*Portugal (Diaz and Azavedo 2004): for sunflower 0.6 wet mass residue/main product ratio (wet assuming 10% moisture). For rape and soy the ratio is 0.8 (13% moisture).

7.4.8 Nut Peelings (Walnut, Almond, and Hazelnut)

These are secondary residues from the peeling plant for nuts.

Calculation of the secondary residues from this source can be done based on the total area (hectares) of nut trees and the average per hectare yield of the main product, and then apply a residue/main product ratio.

$$\text{Area} * \text{per hectare yield of nuts} * \text{residue-to-main product ratio}$$

*Greece (Nikolaou et al., 2002): 0.95–1.5 wet mass residue/main product ratio (wet assuming 5%–8% moisture).

7.4.9 Pressed Grapes Dregs

These are secondary residues from the vine industry. Of the processed grapes 4.6% consists of dregs and 1.5% of stalks (FABbiogas, Italian country report) (http://www.fabbiogas.eu/fileadmin/user_upload/Download/National_Report-Italy-english.pdf).

Calculation of the secondary residues from this source can be done based on the total area (hectares) of vineyards and the average per hectare yield of the main product, and then apply a residue/main product ratio.

Area * per hectare yield of grapes * residue-to-main product ratio

*Portugal (Diaz and Azavedo, 2004): 0.19 wet mass residue/main product ratio (wet assuming 80% moisture).

7.4.10 Fruit Juice Dregs (Oranges and Other Citrus Trees)

These are secondary residues from the juice industry.

Calculation of the secondary residues from this source can be done based on the total area (hectares) of citrus (going to fruit juice production!) and the average per hectare yield of the main product, and then apply a residue/main product ratio.

Area * per hectare yield of citrus fruits * residue-to-main product ratio

*Portugal (Diaz and Azavedo, 2004): 0.56 wet mass residue/main product ratio (wet assuming 80% moisture).

7.4.11 Sugarbeet Industry Residues (Pulp and Molasses)

These are residues from the sugar mill.

Calculation of the secondary residues from this source can be done according to the total area (hectares) of sugarbeet and the average per hectare yield of the main product, and then apply a residue/main product ratio. About 76% of the sugarbeet ends up as pulp in the sugar mill process.

7.4.12 Potato Industry Residues

These are residues from the potato processing industry.

Calculation of the secondary residues from this source can be done based on the total area (hectares) of potatoes converted to starch and feed the average per hectare yield of the main product, and then apply a residue/main product ratio. Main residues concern potato peels and press fiber. About 15% of the processed potatoes (toward chips, crisps, and starch) consist of peels.

The extraction of starch from starch potatoes into potato flour delivers press fiber which is estimated to amount to 15% of the input (e.g., 150 kg/tonne potato input). The press fiber consists for 83.5% of moisture (Elbersen et al., 2011).

7.5 CHALLENGES AFFECTING THE MOBILIZATION OF AGRICULTURAL BIOMASS RESIDUES

The main challenges affecting the mobilization of agricultural biomass residues are:

- understanding *land use* and how it can be sustainably managed and improved;
- *market interactions* to minimize competition and improve synergies and complementarity;
- impacts of *climate change* to the future potentials of agricultural residues;
- *agricultural management and lifestyle* will remain diverse across European regions.

7.6 LAND USE

Efficiency: Different sectors—food, feed, fiber, chemicals, and energy—compete for land. Therefore, in every assessment and future projection, land should be classified based on its physical, chemical, and regional characteristics. Important issues that should be considered include soil quality and structure, cropping patterns, and traditional agricultural practices.

The use of low fertility, marginal land has also been modeled in several studies, investigating future potential. However, it is worth noting that production in marginal lands should meet both economic and sustainable criteria to become competitive. Currently low-input, low-output plant production is generally not profitable for the farmer and therefore may not provide lower unit cost feedstock for the processor. Moreover, the output is often variable in both quantity and quality. The ultimate products from such a system are likely to carry high unit costs and to severely limit the economic viability of the whole chain. Additionally, because low-input production requires many more land, the unit impact on the environment could often be much greater than from a more intensive system.

Hence, planning efforts should focus on choosing the best available cropping and plantation management solutions for each region and land type. Land use exploitation plans should always be moderated to allow for an adequate quantity of residues to be incorporated to the soil before proceeding to removals for other uses.

7.7 MARKET INTERACTIONS

Agricultural residues offer substantial variety in terms of chemical and physical properties. So far the respective production systems operate in such a way that people exploit only part of their production, what is called "primary" product while they leave unexploited significant "residual" quantities.

The use of residues is expected to maximize the added value of the raw materials and improve the income for the producers (farmers, forest community, etc.).

7.8 CLIMATE CHANGE

Climate change is likely to have a significant impact on both the availability of agricultural residues and the crop types cultivated and their regional distribution. For central and northern Europe, an extension of the growing season in spring and autumn is expected, coupled with higher temperatures during the growing period (EEA, 2006). This could enhance crop productivity in these regions. On the other hand, in south and east Europe, an increased risk of drought could lead to productivity losses and increase the risk of soil erosion and water scarcity (EEA, 2006).

7.9 AGRICULTURAL MANAGEMENT AND LIFESTYLE

The prevailing patterns of agricultural management and lifestyle in terms of inputs, water and resource management, cropping strategies, etc. will determine both availability and long-term productivity of land. Improving the respective components based on the climate, ecology, and market requirements will be critical for the mobilization of agricultural residues in each region.

7.10 CONCLUSIONS

The diversity of landscapes and related economic activities in Europe provides a wide range of potential agricultural feedstocks. However, the methodological approaches for estimating their potentials have reached a very good level of harmonization. As work progresses through new initiatives, the respective indicators for residue to main crop ratios, the technical availability factors, and the sustainable removal rates get more refined for specific regions.

To secure year-round sustainable supply for the different end use markets, their production should be linked to current practices as an adaptation strategy to climate change by tackling key issues such as soil quality and maintenance of SOM and soil carbon, land use, and current practices.

Future assessments should:

- focus on bottom-up approaches and account for the soil, climatic, environmental, and socioeconomic profiles in the understudy regions;
- combine regional soil maps with the potential assessment to facilitate the estimation of sustainable removal rates for each region and ensure that soil quality is protected;
- put emphasis on calculating the technical availability factors considering the crop, management practices, climate, and soil conditions in each region.

REFERENCES

Ash, M.S., 1992. Animal Feeds Compendium. AER-656, Commodity Economics, Division, Economic Research Service, USDA.

Bates, J.E., Watkiss, P., and T.W. Thorpe, 1996. The Environmental Implications of Renewables. An interim report produced for the UK Department of Trade and Industry.

Blasi di, C., Tanzi, V., Lanceta, M., 1997. A study on the production of agricultural residues in Italy. Biomass Bioenergy 12, 321–323.

Diaz, J. and Azavedo, J., 2004. Evaluation of biomass residuals in Portugal Mainland. In: Afgan and Carvahlo (Eds.). New and Renewable Energy Technologies for Sustainable Development.

EEA, 2006. European Environmental Agency. How much bioenergy can Europe produce without harming the environment?

Elbersen, W., Janssen, B., Koppejan, J., 2011. De beschikbaarheid van biomassa voor energie in de agro-industrie. Food and Biobased Research, Wageningen UR.

GAO, 1977. To Protect Tomorrow's Food Supply, Soil Conservation Needs Priority Attention. Report CED-77-30. General Accounting Office, Washington, DC.

Hassan, E.G., Alkareem, A.M.A., Mustafa, A.M.I., 2008. Effect of fermentation and particle size of wheat bran on the antinutritional factors and bread quality. Pak. J. Nutr. 7 (4), 521–526.

Larson, W.E., 1981. Protecting the soil resource base. J. Soil Water Conserv. 36, 13–16.

Kent, N.L., Evers, A.D., 1994. Technology of Cereals: An Introduction for Students of Food Science and Agriculture. Woodhead Publishing, 334 p.

Nikolaou, A., Lychnaras, V. and C. Panoutsou. 2002: Characteristics and geographical distribution of agricultural residues for energy production in Greece. 12th European Conference and Technology Exhibition Biomass for Energy and Industry, Amsterdam, June 2002.

Oveson., M.M., 1966. Conservation of soil nitrogen in a summer fallow farming practice. Agron. J 58, 444–447.

Prikhodko, D., Rybchynsky, R., 2009. Wheat Flour: Agribusiness Handbook. Investment Centre Division, FAO, Roma.

Swanson, E.R., Harshbarger, C.E., 1964. An economic analysis of effects of soil loss on crop yields. J. Soil Water Conserv. 19, 183–186.

USDA, 1980. An Analysis of the Timber Situation in the United States: 1952–2030. Forest Service, USDA, Washington, DC.

USDA, 1981. Soil, Water, and Related Resources in the United States. Status, Condition and Trends. 1980 Appraisal Part II. Soil and Water Resources Conservation Act. Department of Agriculture, Washington, DC.

Wieringa, S., 1995. Subversive Women: Women's Movements in Africa, Asia, Latin America and the Caribbean. Zed, London.

WMC, 2008. Wheat Marketing Center. Wheat and Flour Testing Methods: A Guide to Understanding Wheat and Flour Quality, Version 2. Kansas State University Agricultural Experiment Station and Cooperative Extension Service, Kansas State University, USA.

FURTHER READING

Alberson, D.M., Hurst, W.M., 1964. Composting Cotton Gin Waste. Agriculture Research Service, USDA, Washington DC.

Bochner, 1984. Environmental and social costs of biomass energy. BioScience 34 (2), 89–94.

DEA (Danish Energy Agency), 1992. Energy Statistics. Ministry of Energy. DK.

Dos Santos Mathias, T., Moretzsohn de Mello, P., Camporese Servulo, E., 2014. Solid wastes in brewing process: a review. J. Brew. Distill. Vol. 5 (1), 1–9.

Elbersen, B.S., Fritsche, U.R., Petersen, J.-E., Lesschen, J.-P., Böttcher, H., Overmars, K., 2013. Assessing the effect of stricter sustainability criteria on EU biomass crop potential. Biofuels Bioproducts Biorefining (Biofpr) J. (article first published online: 19 Mar 2013), doi:10.1002/bbb.1396.

Elbersen, B.S. Staritsky, 2012. Spatially detailed and quantified overview of EU biomass potential taking into account the main criteria determining biomass availability from different sources. Report for Task 3 in Biomass Futures project. IEE 08 653 SI2. 529 241.

Geletukha, G.G., Zheliezna, T.A., Kucheruk, P.P., Oleynik, Y.M., Tryboi, O.V., 2015. Bioenergy in Ukraine: State of the art and prospects for development. Part 1 Indus. Heat Eng. 37 (2), 68–76.

Jenkins, B.M., Sumner, H.M., 1986. Harvesting and handling agricultural residues for energy. Transact. ASAE 29 (3), 824–836.

Kallivrousis, L., Rozakis, S.T., Vassilatos, V., Mellides, V., Petouni, D., and S., Kyritsis. 1996. Crop residues as a source for decentralised heat and power production in rural areas. The case study of Thrace. In: 9th European Bioenergy Conference. Copenhangen, Denmark, June 24–27 (in press).

Larson, W.E., Pierce, F.J., Dowdy, R.H., 1983. The threat of soil erosion to long-term crop production. Science 219, 458–465.

Majer, S., K. Oehmichen, and A. Brosowski, 2014. Are We Overestimating the Potentials of Biofuels from Wastes and Residues?—Summary of Two Recent DBFZ Studies. Presentation presented at the Fuels of the Future Conference, Berlin, January 20.

Nikolaisen, L., Nielsen, C., Larsen, G.,.M., and P., Keller. 1992. Straw for Energy Production. Technology–Environment–Economy. DEA. DK.

Panagos, P., 2006. The European Soil Database. GEO: connexion 5.

Panagos, P., Liedekerke Van, M., Jones, A., Montanarella., L., 2012. European soil data centre: response to European policy support and public data requirements. Land Use Policy 29 (2), 329–338. Available from: http://dx.doi.org/10.1016/j.landusepol.2011.07.003.

Panos, D.A., 1958. . Effect of wheat straw burning on soil fertility, especially for the conditions of Larissa plain. Athens Acad. Proc. 1957 32, 159.

Pendleton, A.M., and V.P., Moore, 1967. Ginning cotton to preserve fiber quality. ESC-560, Federal Extension Service, United States Department of Agriculture. 19 pp.

Pimentel, D., Fried, C., Olson, L., Schmidt, S., Wagner-Johnson, K., Westman, A., et al., 1984. Environmental and social costs of biomass energy. BioScience 34 (2), 89–94.

Powlson, D.S., Glendining, M.J., Coleman, K., Whitmore, A.P., 2011. Implications for soil properties of removing cereal straw: results from long-term studies. Agron. J. 103, 279–287.

Pudelko, R., Borzecka-Walker, M. & Faber, A., 2013. The feedstock potential assessment for EU-27 + Switzerland in NUTS-3. D 1.2. BioBoost. Project co-funded by the European Commission FP7 Directorate-General for Transport and Energy Grant No. 282873.

Requillart, V., 1984. Upgrading straw in Europe. Biomass News Int., 1, 10-11.

Scarlat, N., Martinov, M., Dallemand, J.F., 2010. Assessment of the availability of agricultural crop residues in the European Union: potential and limitations for bioenergy use. Waste Manage. 30 (2010), 1889–1897.

Spöttle, M., Alberici, S., Toop, G., Peters, D., Gamba, L., Ping, S., et al. (2013). Low ILUC potential of wastes and residues for biofuels. Straw, forestry residues, UCO, corn cobs. Project No. BIEDE13386/BIENL12798.

Summer, H.R., Hellwig, R.E., Monroe, G.E., 1984. Harvesting cotton plant residue for fuel. Transact. ASAE968–972.

Thrän, D., and D. Pfeiffer. 2011. Fokusheft Landschaftspflege. 01/2011 ed. Programmbegleitung des BMU-Förderprogramms "Energetische Biomassenutzung". Leipzig: DBFZ.

Thrän, D., Marcel B., Katja B., Ulrike S., and Vanessa Z., 2010. Globale und regionale Verteilung von Biomassepotenzialen: Status-quo und Möglichkeiten der Präzisierung. BMVBS-Online-Publikation, Nr. 27/2010. DBFZ im Auftrag Bundesministerium für Verkehr, Bau und Stadtentwicklung.

Geletukha G.G., Zheliezna T.A., 2014. Prospects for the use of agricultural residues for energy production in Ukraine. Position Paper N7 by Bioenergy Association of Ukraine (UABio).

World Bank, 2016. Glavonjic, B., Perakis, Ch., Stojadinovic, D. Sector study on biomass-based heating in the Western Balkans. Task 1 report. Analysis of the biomass supply potential.

Chapter 8

Assessing the Potential From Bio-waste and Postconsumer Wood

Martijn Vis

8.1 INTRODUCTION

Waste biomass, also called postconsumer waste, or tertiary residues, is a sustainable source of energy and materials; its utilization has low or no (indirect) land use impacts and associated greenhouse gas emissions. The energy utilization of waste biomass is a form of cascading use, as the biomass has been processed into a product and the resulting waste biomass is used at least once more either for material or for energy purposes (Vis et al., 2016). The utilization of waste biomass for the production of materials can be regarded as multistage cascading, having two or more material uses before final energy utilization. Resource efficient utilization of waste biomass is an important way to bring the principles of the circular economy in practice. A circular economy is one that is restorative and regenerative by design, and which aims to keep products, components, and materials at their highest utility and value at all times, distinguishing between technical and biological cycles (Ellen MacArthur Foundation, 2015). Insight in the potential and quality of available resources is crucial for efficient utilization of the different types of waste biomass, and to obtain a realistic view on their current and future possibilities and limitations.

Waste biomass is a broad concept requiring further specification. Sometimes, secondary residues from wood and agricultural biomass processing have legally the waste status, as the Waste Framework Directive (2008/98/EC) (European Parliament and the Council, 2008) defines waste as any substance or object which the holder discards or intends or is required to discard. However, this chapter focuses on tertiary biomass, released after the use phase of products. More information on secondary forest residues and agricultural residues can be found in Chapter 6, Assessing Lignocellulosic Biomass Potentials From Forests and Industry, and Chapter 7, Assessing Potentials for

Modeling and Optimization of Biomass Supply Chains.
DOI: http://dx.doi.org/10.1016/B978-0-12-812303-4.00008-2

Agricultural Residues, respectively. Important types of tertiary waste biomass are bio-waste, postconsumer wood, waste paper, and sewage sludge. In this chapter, bio-waste and postconsumer wood have been selected for further investigation, because of their potential and possible applications. Waste paper is a very relevant resource as well, but mainly recycled for paper production and less an option for energy and other material applications.

Bio-waste consists of biodegradable garden and park waste, food and kitchen waste from households, restaurants, catering, and retail premises, and comparable waste from food processing plants. It can be part of integrally collected municipal waste and used for energy generation by incineration with the other waste, or separated after collection and used for both refuse derived fuel production and anaerobic digestion. Separate collection of bio-waste, for instance kitchen and garden waste, is available for composting, which can be combined with anaerobic digestion.

Postconsumer wood includes all kinds of wooden material that is available at the end of its use as a wooden product, like packaging materials (e.g., pallets), demolition wood, timber from building sites, and used furniture. The quality of the postconsumer wood determines the possibilities to utilize postconsumer wood for material applications beyond combustion with energy application. The quality of postconsumer wood depends on its former application, whether the material is painted, impregnated, or otherwise treated; whether it consists of sawn wood or panels; whether it is glued, nailed, or otherwise stuck together with other materials; if it is collected separately or integrally with other waste, etc. (Vis et al., 2016). Fig. 8.1 shows an impression of the material applications of postconsumer wood.

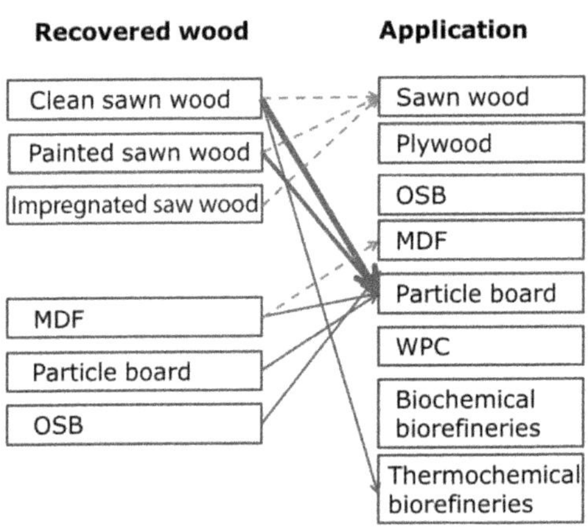

FIGURE 8.1 Material applications for different types of postconsumer wood. Big arrows represent the main applications, small arrows common applications, but smaller in volume; dotted arrows show possible future applications, with currently limited commercial application. *Vis, M., 2016. Barriers to cascading use, presented at the 6th ISG Meeting of the Study on the Optimised Cascading Use of Wood, Brussels.*

Section 8.2 discusses definition, key challenges, a straightforward method for the assessment of the availability of bio-waste, and some results of the application of this method. In Section 8.3, the same is provided on postconsumer wood, while in addition attention is paid to the possible classification in quality classes of postconsumer wood. Resulting resource potentials at NUTS3 level for almost all European countries can be found in the S2Biom toolset available on www.s2biom.eu. In Section 8.4 conclusions and recommendations are drawn.

8.2 BIOMASS RESOURCE ASSESSMENTS OF BIO-WASTE

8.2.1 Definitions

More than secondary biomass residues, the treatment of tertiary waste biomass is subject to European and national legislation, such as the Waste Framework Directive and national waste management plans. For this reason, it makes sense to define waste biomass in line with relevant European legislation such as the Landfill Directive (1999/31/EC) (European Council, 1999) and the Waste Framework Directive (2008/98/EC) (European Parliament and the Council, 2008). The oldest definition of waste biomass is the found in the Landfill Directive. It defines biodegradable waste as *"any waste that is capable of undergoing anaerobic or aerobic decomposition, such as food and garden waste, and paper and cardboard."* Despite that the amount biodegradable municipal waste produced in the EU Member States is highly relevant to verify whether the targets of the Landfill Directive are met, the amounts cannot be found directly in Eurostat. Vis and Dees (2011) provide a relatively simple method to estimate the amount of biodegradable municipal waste produced on country level. Almost 10 years later, the Waste Framework Directive was published defining bio-waste as *"biodegradable garden and park waste, food and kitchen waste from households, restaurants, catering and retail premises and comparable waste from food processing plants."* Bio-waste is part of biodegradable municipal waste as defined in the Landfill Directive, but it excludes other fractions included in biodegradable municipal waste such as textile, paper and paperboard, wood, and sewage sludge. Textile and separately collected paper and paper board have existing material applications such as paper recycling, and thus not available for bioenergy production and new material applications without disturbing existing recycling loops. Waste wood and sewage sludge are relevant but because of the different composition and applications these should be analyzed separately. Therefore, the resource assessment method presented in this chapter will focus on the availability of bio-waste as defined in the Waste Framework Directive.

8.2.2 Key Challenges

Assessing the availability of bio-waste is challenging mainly because it is not only found in separately collected as kitchen and garden waste, but it is also part of integrally collected municipal waste. In order to estimate the bio-waste fraction of municipal waste, an up-to-date waste composition analysis of the integrally collected fraction needs to be applied. Within one country, the bio-waste fraction of municipal solid waste (MSW) will vary between regions and seasons, information which is usually not available in resource assessments or statistics on bio-waste. In fact, a large number of Member States does not collect specific data on the quantities of bio-waste and it cannot be found as such in Eurostat. Franckx et al. (2010, p. 49) point to the fact that not all Member States use the same terminology and collect data on different fractions of biodegradable waste, which do not exactly match with the European definition of bio-waste. Furthermore, several Member States do not provide data on the organization and coverage of separate bio-waste collection. Moreover, the collection and treatment of kitchen waste depends on the implementation of the Animal By-Products Regulation (1069/2009) (European Parliament and the Council, 2009), which can vary even within countries. The proposed method makes use of bio-waste fractions of MSW as established in Franckx et al. (2010), but as this study is not regularly updated, these fractions need to be updated at a certain point.

There are a specific challenges related to obtaining spatially explicit bio-waste potentials. Bio-waste does not arise from larger areas like forests and agricultural crops, but is released by households and transported to waste collection points. This means that it is not possible to obtain site-specific information by, for example, remote sensing. Site-level information on bio-waste has to be collected site-by-site, and cannot be automated like in case of agricultural biomass like wheat straw. If this site level of detail is not required, for instance in resource assessments covering a large area of one or more countries, it is possible to obtain bio-waste volumes at regional level (NUTS3). (NUTS is the Nomenclature of Territorial Units for Statistics; NUTS0 is country level, NUTS3 is the level of counties, regions, a cluster of municipalities, depending on the country of concern.) In all countries data on MSW is available in kg per capita. Combining this with the bio-waste fraction and the population at NUTS3 level, results in an estimate of bio-waste availability at NUTS3 level. This method, however, ignores differences in the production of bio-waste between urban and rural areas, and between regions within the country in general. Also a fixed factor between bio-waste and total municipal waste is applied as known at country level.

Dependency on conversion factors and export opinion is another challenge worth mentioning. In the translation of mass to energy potentials of bio-waste, it is important to distinguish between separately collected and integrally collected bio-waste, as these have different possibilities for energy application, e.g., bio-waste as integral part of municipal waste can be combusted, and separately collected bio-waste can be digested before composting. These fractions have been estimated by Franckx et al. (2010), but will require future updating. In order to convert volumes to energy potentials, conversion factors need to be applied, which are unavoidably based on the experts' selection of literature. Possible fluctuations in moisture content and biogas yield of the waste can easily introduce a substantial error (say 10%–15%). The conversion factors between volumes and energy content can be improved if more data would be available on the moisture content, composition, and potential biogas yield of bio-waste. Given the varying composition of bio-waste between regions and seasons, this would require substantial effort.

8.2.3 Proposed Methodology

In the absence of European statistical information on bio-waste production, the proposed methodology provides a cost and time efficient way to estimate the availability of bio-waste in larger areas at NUTS3 level. The method was developed within the S2Biom project, building on a standardized method for biodegradable municipal waste production as developed in the Bio Energy Europe Project (BEE) (Vis and Dees, 2011). The proposed method uses data on the bio-waste fractions of municipal waste from Franckx et al. (2009, 2010), a recent EU study dedicated to bio-waste, that can be regarded as the most extensive and complete European-wide study on bio-waste availability. Other European studies include bio-waste or biodegradable municipal waste as part of work covering a broader range of biomass types. The BEE project (BEE, 2010) provides an interesting overview of methods to determine biodegradable waste as applied in European biomass resource assessments, such as Siemons et al. (2004), EEA (2006), and Thrän et al. (2006). The studies either developed assumptions on the biodegradable fraction of MSW and linked them to MSW statistics available at Eurostat, or simply cited other studies. The projections in these studies differ somewhat from each other in assumptions on the expected growth of bio-waste production (linked to GDP and/or population growth), and in general take into account the impact of the Landfill Directive, i.e., the demanded shift of biodegradable waste from landfill to other treatment methods like incineration, digestion, and composting. The biodegradable waste data in the more recent Biomass Futures project (Elbersen et al., 2012) is taken from the earlier EEA (2006) study, which works with data provided by Skovgaard and Moll (2005) and own assumptions on future waste generation. The method proposed below contains, where possible, a country-specific bio-waste to MSW fraction as developed by Franckx et al. (2009), assumes in its projections that the Landfill Directive will be achieved by all Member States, and that bio-waste generation will follow the population growth. Moreover, like in Siemons et al.

(2004), a distinction is made between separately collected bio-waste and bio-waste as part of integrally collected MSW, as both options have different energy generation options.

To summarize, the availability of bio-waste in year x on NUTS3 level can be established as MSW generated per capita (kg/capita) times the bio-waste fraction (%) times the population of the NUTS3 area (persons). A further distinction is made between the separately collected bio-waste and bio-waste as part of mixed waste, as a basis to estimate the energy potentials of the bio-waste. The different elements are further elaborated as follows.

8.2.3.1 MSW per Capita

European statistics provide information on the amounts of MSW generated per capita in a country. [See Eurostat: municipal waste generation and treatment, by type of treatment method (tsdpc240)]. The data is available at country level (NUTS0). It is likely that differences exist in quantities of MSW per capita between regions and that the composition differs between urban and rural areas, etc. Eurostat has carried out a pilot on collection data on municipal waste per capita by NUTS2 regions. [See Eurostat: municipal waste by NUTS2 regions—pilot project (env_rwas_gen)]. The project covered, however, a limited number of countries and data collection has stopped after 2011. Therefore, for more recent years, country-level MSW per capita data will need to be used.

8.2.3.2 Bio-waste Fraction

Eurostat does not publish data on the share of bio-waste within MSW. This ratio has to be collected from statistical information and sorting analyses on national level. Franckx et al. (2009) have analyzed literature on the share of bio-waste in MSW in all the EU27 countries. In case no data could be found for a particular country, the study used the share of biodegradable municipal waste in municipal waste that is known for the year 1995 because of the implementation of the Landfill Directive, multiplied with a factor of 56% bio-waste in biodegradable municipal waste. The latter factor is based on the composition of household waste in Pleven and Flanders and the assumption that total biodegradable waste consists of bio-waste + paper + textiles + ½ of other fractions. The bio-waste fractions established by Franckx et al. (2009) are recommended as they form the most up-to-date complete set of bio-waste fractions for the EU27 currently available. For Croatia (the 28th EU country), it is assumed that the bio-waste fraction is the average fraction of neighboring countries Slovenia and Hungary. For the non-EU countries, no data on the bio-waste fraction has been collected; instead, the average bio-waste fraction of 35.9% as established by Franckx et al. (2009) has been used.

8.2.3.3 Population Data

Historic population data on NUTS3 level can be taken from Eurostat. [Eurostat: population change—demographic balance and crude rates at regional level (NUTS3) (demo_r_gind3)]. Projections on the development of the total quantity of bio-waste are assumed to be proportional to population growth. The main scenario on population development from Eurostat can be used to predict the population in 2020. [Eurostat: main scenario—population on 1 January by age and sex (proj_13npms)]. This information is only available on country level (NUTS0). In order to establish population data in 2020 and 2030, the NUTS3-level population data of year *x* needs to be multiplied with the expected change in population in 2020 and 2030 (or any other year in the future) compared to year *x* on national level.

The development of bio-waste availability has not been linked to GDP growth, given the uncertainty of GDP development and the fact that many EU countries will reach or have reached decoupling with GDP. This is a conservative assumption, especially for European countries with still a relatively low GDP.

8.2.3.4 Separately Collected Bio-waste Versus Bio-waste in Mixed Waste

Franckx et al. (2009) have analyzed the percentages of bio-waste that is collected separately or exist as part of mixed waste for the EU27 on country level. These numbers (base year 2008) are still the best currently available. In Franckx et al. (2009) also projections have been provided of the shares of bio-waste going to the different treatment options like landfill, incineration, MBT (mechanical biological treatment), composting, backyard composting, anaerobic digestion, and others have been made for the years 2008–20. It has been assumed that all countries meet the requirement of the landfill directive, e.g., that maximally 35% of the amount of biodegradable waste generated in base year 1995 is landfilled in 2020, even if current developments show that diversion from landfill has not been successful yet. Furthermore, the projections are based on policy views and changes in treatment of bio-waste in the member state concerned. For instance, some countries have a strong preference for MBT, others for incineration with energy recovery. For the year 2030, the same shares between treatment options are used as in the year 2020, as currently no policies are known that influence the production of bio-waste after 2020.

8.2.3.5 Conversion Factors

Table 8.1 provides conversion factors that can be used to convert the mass to volumes and energy. The data has been retrieved from the biomass properties database developed in the S2Biom project (Annevelink

TABLE 8.1 Conversion Factors to Convert Mass Potentials of Bio-waste to Energy Potentials

	NVC_{ar} (MJ/kg)	Moisture Content ($w\%_{ar}$)	Basic Density (kg/m^3)
Bio-waste as part of integrally collected municipal waste: biodegradable waste of not separately collected municipal waste (excluding textile and paper)	10.8	27.2%	500
Separately collected bio-waste: biodegradable waste of separately collected municipal waste (excluding textile and paper)	4.3	55.6%	500

et al., 2016) (see http://www.s2biom.eu/images/Publications/Annex_D2.4_S2BIOM_biomass_properties.pdf).

8.2.4 Results

Fig. 8.2 shows the production of bio-waste per year in the different European countries. The expected growth in bio-waste production is linked to the population growth in these countries and is rather stable, i.e., from 108 million tonnes in 2012 to 109 and 111 million tonnes in 2020 and 2030, respectively. In 2012 about 23 million tonnes (21%) of the bio-waste was separately collected, and is expected to increase to 39 million tonnes in 2020 (36%) and 43 million tonnes (39%) in 2030. Detailed datasets and figures can be found in the toolset available on the website of S2Biom (go to http://www.s2biom.eu/en/, click on "toolset" and login to http://s2biom.alterra.wur.nl/).

8.2.4.1 Regional Disaggregation

The data on bio-waste has been collected on country level. The disaggregation of the results at NUTS3 (municipal) level has been performed based on the population of these regions, as population is the best indicator for waste generation. Consequently, the biomass densities in the maps show a strong correlation with the population density in the different areas of Europe (Fig. 8.3).

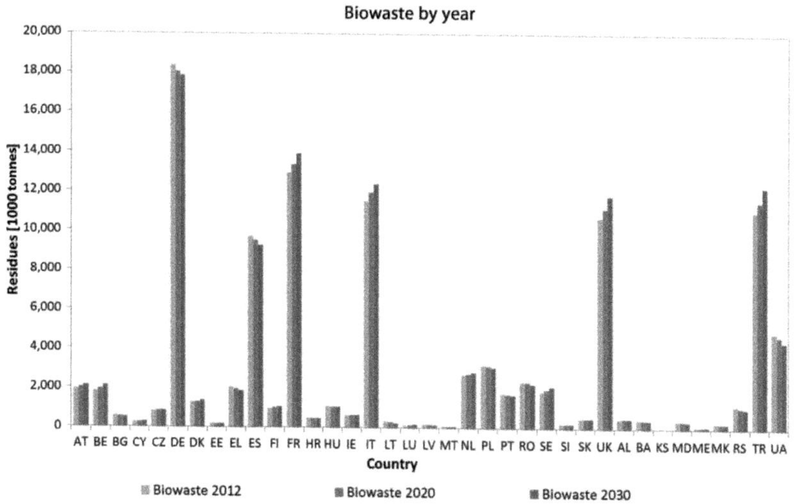

FIGURE 8.2 Bio-waste production in 2012 and projections of 2020 and 2030. *Dees, M., Datta, P., Höhl, M., Fitzgerald, J., Verkerk, H., Zudin, S., et al., 2017. Atlas with regional cost supply biomass potentials, Version 1.1. Deliverable D1.4 of the S2Biom project, grant agreement no. 608622.*

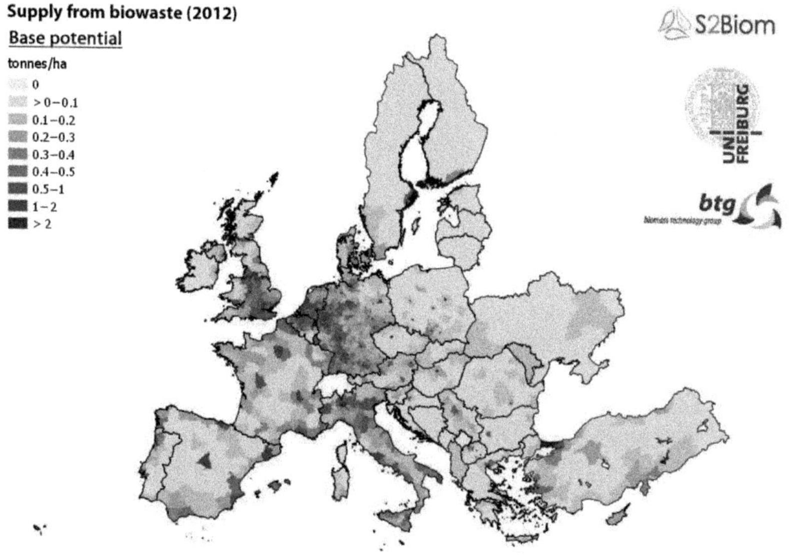

FIGURE 8.3 Regional distribution of the production of bio-waste production in 2012. *Dees, M., Datta, P., Höhl, M., Fitzgerald, J., Verkerk, H., Zudin, S., et al., 2017. Atlas with regional cost supply biomass potentials, Version 1.1. Deliverable D1.4 of the S2Biom project, grant agreement no. 608622.*

8.3 BIOMASS RESOURCE ASSESSMENTS OF POSTCONSUMER WOOD

8.3.1 Definitions

Postconsumer wood includes all kinds of wooden material that is available at the end of its use as a wooden product ("postconsumer" or "postuse" wood). Postconsumer wood mainly comprises packaging materials (e.g., pallets), demolition wood, timber from building sites, as well as fractions of used wood from residential (municipal waste), industrial, and commercial activities (e.g., used wooden furniture). A further distinction between hazardous and nonhazardous fractions, or a more refined categorization in waste classes (e.g., A: unpainted and untreated wood; B: painted and glued wood; C: treated wood with, for example, creosotes or preservatives) can be useful for analysis of possible bioenergy and new material applications.

There are no European legal definitions of postconsumer wood and further classification. Instead, definitions as found in literature can be followed. According to the COST Action 31 (Jungmeier, 2005), recovered wood includes all kinds of wooden material that is available at the end of its use as a wooden product ("postconsumer" or "postuse" wood). In the EUwood project (Leek, 2010), a slightly different definition is used: *Postconsumer wood (PCW) includes all kinds of wooden material that is available at the end of its use as a wooden product ("postconsumer" or "postuse" wood).* The latter definition is used in the proposed biomass resource method description in the next sections.

8.3.2 Key Challenges

It is difficult to obtain reliable estimations of postconsumer wood available for energy and material applications. With the exception of wooden packaging waste [Eurostat: packaging waste by operations and waste flow (env_waspac)], data on postconsumer wood cannot be obtained directly from Eurostat, meaning that in practice resource assessments depend on study reports that rely on sector information and expert opinion. Eurostat provides information on "wood waste" [Eurostat: generation of waste by waste category, hazardousness, and NACE Rev. 2 activity (env_wasgen)], but this is a mixture of several types of secondary wood processing wood and tertiary postconsumer wood. Wood waste (07.5) consists of wooden packaging (07.51) sawdust and shavings (07.52), waste bark, wood waste from mechanical waste treatment (Eural 19 12 07), separate collected waste wood (Eural 20 01 38), and the hazardous fractions thereof (European Commission, 2004). Because only data on the main category wood waste is available, Eurostat data cannot be used to determine the potential of postconsumer wood.

Another challenge is the lack of uniform quality classes for postconsumer wood fractions applied in the different EU Member States (Vis et al., 2016). Postconsumer wood exists in various quality classes that indicate their possibilities for material reuse and energy application. One important distinction is between hazardous and nonhazardous waste wood, as made in European statistics. In general, hazardous wood cannot be used in most material applications or combusted for energy generation without special measures, while more opportunities exist for the nonhazardous fraction. Yet, there is a need for a more detailed classification of postconsumer wood fractions. Several countries apply more refined classification methods, like Germany (AltholzV, 2002), Austria (RecyclingholzV, 2012), United Kingdom (PAS 111, 2012), Flanders (VLAREM, 1991), and the Netherlands (LAP II, 2014). As an example, the Dutch waste management plan distinguishes three main categories of postconsumer wood: A-quality, unpainted and untreated wood; B-quality, wood not mentioned under A-quality, among others, painted, lacquered and glued wood; and C-quality, treated wood. A-quality wood can be recycled or used for material recycling. B-quality wood can be used for both applications as well, given that certain treatment is provided (removing paint) or emission reduction equipment. A- and B-quality woods are often provided as mixtures, therefore it is not possible to distinguish between both categories in statistics. Both qualities will be indicated as nonhazardous wood. C-quality (hazardous) consists of treated wood like wood treated with creosotes, wood treated with wood preservatives containing copper, chrome, and arsenic (CC and CCA wood), and wood treated with other means (fungicides, insecticides, etc.). C-wood is a distinct category, in general not suitable for material recycling (with the exception of material reuse of creosoted wood), but in general this wood can be combusted for energy generation, provided that sufficient measures are taken, especially advanced emission reduction measures. Other countries have similar but slightly different quality classifications, which makes harmonization of resource potentials on basis of quality, beyond a simple classification into hazardous and nonhazardous postconsumer wood difficult to implement.

Furthermore, it is particularly difficult to obtain spatially explicit postconsumer wood potentials. Like bio-waste, postconsumer wood does not arise from larger areas like forests and agricultural crops, but is released from various sources, like construction and demolition activities, kerbside consumer waste, and from other sources. Data on national level can in principle be disaggregated to regional (NUTS1–3) level, based on population distribution, but the actual resource depends on, among others, the locations of building and demolition activities, which can vary between regions. As demolition and construction activities are expected to fluctuate more in location and volume than the production of bio-waste, disaggregation to regional level can create substantial estimation errors.

The issue of competing uses plays a role in the estimation of the resource potential of postconsumer wood for bioenergy vs existing and new material applications. Currently, mainly the particle board industry is able to utilize part of especially the cleaner sawn wood fractions of postconsumer wood such as packaging wood but also smaller fractions of used panels (Vis et al., 2016). Within the S2Biom project (Dees et al., 2016), three scenarios were developed for the resource potential resulting in (1) a technical potential covering all produced postconsumer wood, (2) a base potential taking into account the current use of postconsumer wood in material applications such as particle board, and (3) a "circular economy" potential scenario, in which targets of proposed legislation on the circular economy and wood quality determine the availability of postconsumer wood for energy. One point of concern is that in some regions (for instance Flanders), competition of energy and material sectors for postconsumer wood is a big issue, while in other regions (Nordic countries), this is not the case.

Finally, it is worth mentioning that due to the limited availability of statistics and in depth European studies on availability of postconsumer wood, there is a lack of primary data collection on this topic, and a high dependency on expert judgment and a limited number of previous studies. Moreover, in order to go from volumes to energy potentials, conversion factors are introduced, which are based on the experts' selection of literature, and it can easily introduce an error in the order of 10%–15%.

8.3.3 Proposed Methodology

Given that, with the exception of used packaging wood (Eurostat: generation of waste by waste category, hazardousness, and NACE Rev. 2 activity (env_wasgen)), Eurostat data cannot be used to determine the potential of postconsumer wood, data needs to be collected from literature, expert judgment, and/or from primary research. The latter is very time and cost intensive. No studies with European coverage that solely focused on postconsumer wood were identified. BEE (2010) provides an overview of methods used to obtain data on postconsumer wood in a number of European biomass resource assessments: Siemons et al. (2004), Thrän et al. (2006), and EEA (2006). Siemons et al. (2004) based its estimates on Brodersen et al. (2002), which presents the results of a questionnaire from the European Environment Agency (EEA) on the availability of demolition wood in seven European countries. Thrän et al. (2006) used a figure on waste wood production per capita in Germany to estimate its availability in all European countries. EEA (2006) estimated availability of construction and demolition wood being "wood offcuts from building construction and wood recovered during demolition." In order to project changes in the amount of construction and demolition wood, the EEA used the projections of construction and demolition waste (not only wood) made by the European

Topic Centre on Resource and Wastes Management (Skovgaard and Moll, 2005; EEA, 2005). It is not clear how the EEA obtained base year data on demolition wood, since the referred studies only cover demolition waste and do not distinguish the wood fraction. The future projections of waste wood are related to the economic activity in the construction and demolition sector.

The proposed method as developed and applied in the S2Biom project is based on literature data from the more recent EUwood (Mantau et al., 2010; Leek, 2010) and EFSOS II studies (UNECE and FAO, 2011). EUwood combines, among others, Eurostat and COST E31 (2007) data. The EFSOS II data on postconsumer wood is based on EU wood, but covers Europe as a whole instead of EU28. The resource potential of postconsumer wood can be described as follows:

$$PCW_{\text{technical potential}} = PCW_{\text{material}} + PCW_{\text{energy}} + PCW_{\text{disposed}}$$
$$PCW_{\text{sustainable potential}} = PCW_{\text{energy}} + PCW_{\text{disposed}}$$

where:

$PCW_{material}$ = postconsumer wood used for materials like panels and chipboards
PCW_{energy} = postconsumer wood used for energy production
$PCW_{disposed}$ = landfilled and/or incinerated without energy generation with MSW.

In order to determine the sustainable potential of postconsumer wood available for energy, it is necessary to estimate how much is used for material applications. Leek (2010, pp. 110–120) provides a table on the availability of *PCW material* (for material recycling) and *PCW energy* for 2007. Based on information from COST E31 for 17 countries, national experts gave their best professional judgment on how recovered wood is used for the production of panels and for energy purposes, and how much postconsumer is not used, e.g., landfilled or incinerated (Leek, 2010). This information has been extrapolated for use in the proposed method.

In case of estimation of the future sustainable energy potential of postconsumer wood, one could also take into account relevant policy developments. For instance, The Circular Economy Package (European Commission, 2015) proposes a target of 75% of material recycling of packaging wood in 2030; this will be a challenge, but the quality of packaging waste (mainly clean sawn wood) is suitable for recycling. (COM/2015/0596 final—2015/0276 (COD). Proposal for a Directive of the European Parliament and of the Council amending Directive 94/62/EC on packaging and packaging waste.) The other postconsumer wood fractions are more difficult to recycle; there are not too many options to recycle used panels (particle board, MDF, OSB, plywood). Therefore, it can conservatively be assumed that recycling rate of other postconsumer wood (besides packaging wood) is expected not to exceed 50%. Furthermore, given the quality requirements for material applications of wood, all hazardous postconsumer wood can be assumed to be available for energy generation.

8.3.3.1 Distinction Between Hazardous Versus Nonhazardous Wood

Eurostat provides data on the hazardous and nonhazardous fractions of waste, but given the earlier mentioned limitations on Eurostat reporting on postconsumer wood, it will be necessary to collect data on the percentage hazardous wood on country level. For instance, according to Probos (2014) in the Netherlands, yearly around 1000–1400 ktonnes A/B-wood and 80–120 ktonnes/year of hazardous C-wood is produced in the period 2007–12. This means that hazardous C-wood forms on average 7.6% (7.4%–7.8%) of total postconsumer wood supply in the Netherlands. According to a dedicated case study in the Bioxchange project (Benthem et al., 2005), in Germany 17% of the postconsumer wood is hazardous. According to the same study in the Netherlands the share is lower, only 6%.

8.3.3.2 Population Data

Historic population data on NUTS3 level can be taken from Eurostat. [See Eurostat: population change—demographic balance and crude rates at regional level (NUTS3) (demo_r_gind3)]. Projections on the development of the total quantity of postconsumer wood are assumed to be proportional to population growth. The main scenario on population development from Eurostat can be used to predict the population in future years. [See Eurostat: main scenario—population on 1 January by age and sex (proj_13npms)].

8.3.3.3 Conversion Factors

Table 8.2 summarizes conversion factors that can be used to convert mass to volumes and energy. The data has been retrieved from the biomass properties database developed in S2Biom (Annevelink et al., 2016).

TABLE 8.2 Conversion Factors to Convert Mass Potentials of Postconsumer Wood to Energy Potentials

	NVC_{ar} (MJ/kg)	Moisture Content ($w\%_{ar}$)	Basic Density (kg/m^3)
Hazardous postconsumer wood	14.2	13.9%	500
Nonhazardous postconsumer wood	16	13.1	500

8.3.4 Results

Fig. 8.4 shows the production of postconsumer wood in the different European countries available for energy production. This figure excludes the current material application of 13 million tonnes of recovered wood (e.g., for panel board production), and it is assumed that the material application remains constant in 2020 and 2030. It includes all postconsumer wood currently landfilled. The growth in availability from 18 million tonnes in 2012 to 20 and 22 million tonnes in 2020 and 2030, respectively, is related to the population growth.

Fig. 8.5 shows three scenarios for 2030: the technical potential including current material applications of postconsumer wood; the base potential excluding current material applications of wood (which equals the numbers in Fig. 8.4); and a "circular economy" potential that takes into account increased material recycling of used packaging wood to 75% in 2030 as proposed by the European Commission (2015) as part of their Circular Economy package, and material application of 50% of all nonhazardous postconsumer wood. The total technical potential in 2030 is 35 million tonnes, the base potential 22 million tonnes, and with additional cascading use in the circular economy scenario 19 million tonnes.

The difference between the user-defined scenario of increased cascading use of wood and the base scenario base potential depends on the historical shares of energy use, material application, and disposal of postconsumer wood

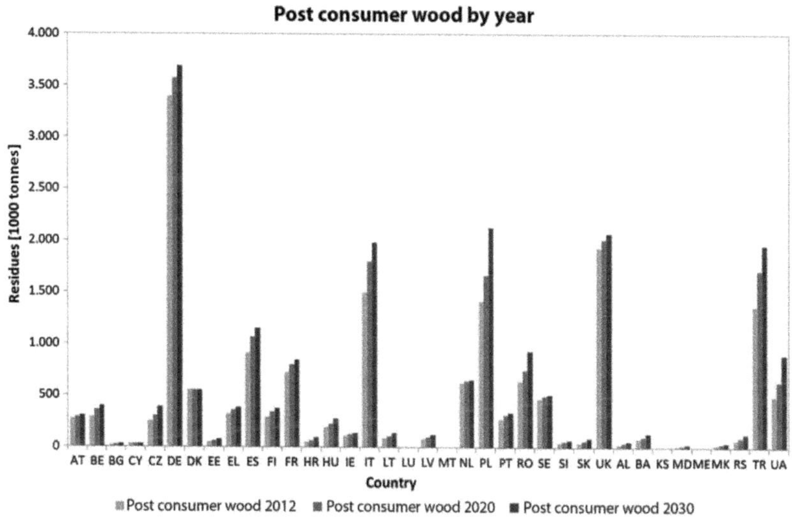

FIGURE 8.4 Postconsumer wood production in 2012 available for energy production and projections for 2020 and 2030, taking into account current material applications of postconsumer wood. *Dees, M., Datta, P., Höhl, M., Fitzgerald, J., Verkerk, H., Zudin, S. et al., 2017. Atlas with regional cost supply biomass potentials, Version 1.1. Deliverable D1.4 of the S2Biom project, grant agreement no. 608622.*

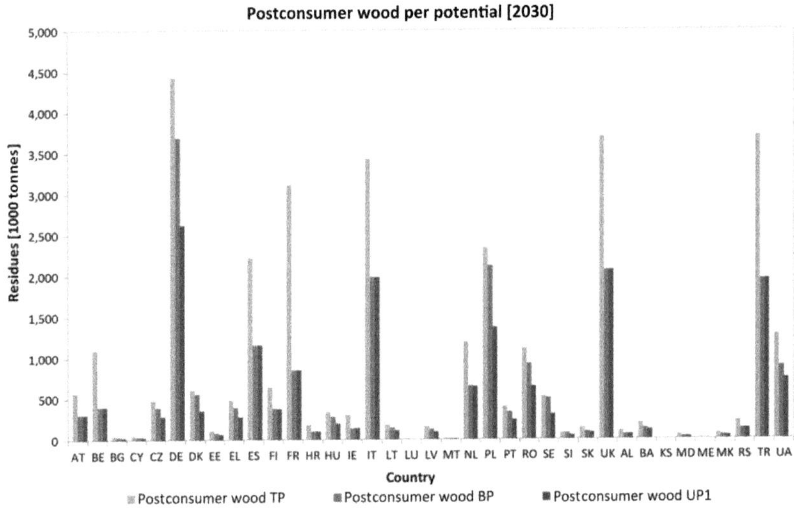

FIGURE 8.5 Technical, base, and circular economy potential of postconsumer wood production for energy in 2030. *Dees, M., Datta, P., Höhl, M., Fitzgerald, J., Verkerk, H., Zudin, S. et al., 2017. Atlas with regional cost supply biomass potentials, Version 1.1. Deliverable D1.4 of the S2Biom project, grant agreement no. 608622.*

on country level. In Denmark, Germany, and Sweden, the circular economy potential is lower than the base potential because these countries have currently a high energy utilization of postconsumer wood, which would be reduced in the user-defined scenario with increased material application of wood in 2020 and 2030. In countries like Czech Republic, Estonia, Greece, Hungary, Latvia, Lithuania, Poland, Portugal, Romania, and Slovenia, currently much wood is still disposed, which is in case of the cascading use scenario will be partly allocated to material application in 2020 and 2030. This results in a lower user-defined potential compared to the base scenario, in which all currently disposed wood is allocated to energy use. In case of a tradition of high material utilization and a low disposal rate, the difference between base potential and user-defined potential is zero. In this case the base potential (that takes into account current material application) meets already the assumptions on increased material consumption of the user-defined potential. This is the case in countries like Austria, France, and Italy.

The detailed datasets and figures can be found in the toolset available on the website of S2Biom (go to http://www.s2biom.eu/en/, click on "toolset" and login to http://s2biom.alterra.wur.nl/).

8.4 CONCLUSIONS

Bio-waste and postconsumer wood are relevant tertiary residues and a sustainable source of energy and materials; its utilization has low or no (indirect) land use impacts and associated greenhouse gas emissions. Bio-waste is defined in the Waste Framework Directive but differences have been found (Franckx et al., 2010) in national interpretation between EU Member States. Postconsumer wood is not yet defined in European legislation. An EU-wide classification of the different postconsumer wood assortments, beyond hazardous and nonhazardous waste, could help to improve the sharing and trade in wood resources across the EU, the understanding of potential end-uses and lead to new market developments (Vis et al., 2016).

Assessment of the resource potential of bio-waste and postconsumer wood is challenging. European statistics present aggregated data on MSW and wood waste production, but obtaining the bio-waste and postconsumer waste fractions requires further estimations by expert judgment or thorough analysis of national information sources, if available. A main step forward would be if Eurostat would directly publish bio-waste data and enabled to present data on the several subcategories of wood waste already defined in Eurostat, including wooden packaging, wood waste from mechanical waste treatment, separate collected waste wood, and the hazardous fractions thereof.

In absence of direct statistical information, the proposed resource assessment methods can help to estimate bio-waste and postconsumer wood volumes in a relatively short time, without too much effort. These methods are partly relying on the best available existing European studies that at some point—say at least each 5 years—need to be updated. This update includes a country-level assessment of bio-waste fractions and type of treatment of the bio-waste, and for postconsumer wood a national-level assessment of postconsumer wood arising and applications.

There are a specific challenges related to obtaining spatially explicit bio-waste and postconsumer wood potentials. These tertiary residues do not arise from larger areas like forests and agricultural crops, but are released by households and transported to waste collection points, or arise at construction and demolition sites. This means that it is not possible to obtain site-specific information by, for example, remote sensing. Site-level information on bio-waste has to be collected site-by-site, and cannot be automated like in case of agricultural biomass like wheat straw. It is, however, possible to obtain an estimation of regional activities by disaggregation of the national data to regional level (NUTS3) by population distribution. It has to be taken into account that this method does not properly take into account regional differences, between, for example, rural and urban areas and possible differences in bio-waste generation.

The application of the proposed resource assessment methods shows that the production of bio-waste in Europe is significant and stable, i.e., increasing from 108 million tonnes in 2012 to 109 and 111 million tonnes in 2020 and 2030, respectively (the EU28 plus Iceland, Norway, Switzerland, Albania, Bosnia-Herzegovina, Serbia, Montenegro, Macedonia, Turkey,

Belarus, Moldavia, Ukraine). Given that separate collection of bio-waste is expected to grow from 21% in 2012 to 39% in 2030, the opportunities for anaerobic digestion in combination with composting are promising. Postconsumer wood arises in relatively smaller volumes but have interesting possibilities for material and (subsequent) energy application. It is estimated that in 2012 in total 31 million tonnes of postconsumer wood is produced, of which 13 million tonnes already applied in existing material applications, leaving 18 million tonnes a potential for energy and new material applications. In case of a circular economy scenario with optimized cascading use of wood, in 2030 about 17 million tonnes of postconsumer wood can be used for material applications and 19 million tonnes for energy generation. The cleaner mainly sawn wood fraction of postconsumer wood is most suitable for material applications, while the hazardous wood fractions and the used panels will be more difficult to process for material use, and left for energy generation.

REFERENCES

Altholz, V., 2002. Verordnung über Anforderungen an die Verwertung und Beseitigung von Altholz (Altholzverordnung—AltholzV). The German waste wood ordinance.

Annevelink, W., Elbersen, E., Alakangas, B,. Elbersen, B., Annevelink, J., Ramirez Almeyda, T., Lammens, 2016. Database for standardized biomass characterization. Deliverable D2.4 of the S2Biom project. Available from: <http://www.s2biom.eu/images/Publications/Annex_D2.4_S2BIOM_biomass_properties.pdf>.

BEE, 2010. Methods & Data Sources for Biomass Resource Assessments for Energy, D.4.3 & D.4.4 of the Biomass Energy Europe (BEE) Project Final.

Benthem, M., van, N., Leek, U., Mantau, H., Weima, 2005. Markets for recovered wood in Europe: Case studies for the Netherlands and Germany based on the Bioxchange project. <http://www.probos.nl/images/pdf/artikelen/PaperProbosforCOST-E31.pdf>.

Brodersen, J., Juul, J., Jacobsen, H., 2002. Review of Selected Waste Streams: Sewage Sludge, Construction and Demolition Waste, Waste Oils, Waste From Coal-Fired Power Plants and Biodegradable Municipal Waste. European Environment Agency, Copenhagen, Denmark.

Cost E31, 2007. 3rd European COST E31 Conference. Management of Recovered Wood. Proceedings, Klagenfurt, Austria.

Dees, M., Elbersen, B., Fitzgerald, J., Vis, M., Anttila, P., Forsell, N., et al., 2016. A spatial data base on sustainable biomass cost–supply of lignocellulosic biomass in Europe—methods & data sources. Project Report. S2Biom—a project funded under the European Union 7th Frame Programme. Grant Agreement no. 608622. University of Freiburg, 170 pp.

Dees, M., Datta, P., Höhl, M., Fitzgerald, J., Verkerk, H., Zudin, S., et al., 2017. Atlas with regional cost supply biomass potentials, Version 1.1. Deliverable D1.4 of the S2Biom project, grant agreement no. 608622.

EEA, 2005. Greenhouse Gas Emission Trends and Projections in Europe 2006. European Environment Agency, Copenhagen, Denmark, p. 68.

EEA, 2006. How Much Bioenergy Can Europe Produce Without Harming the Environment? European Environment Agency (EEA), Copenhagen, Denmark, p. 59. (plus Appendices).

Elbersen, B., Startisky, I., Hengeveld, G., Schelhaas, M.J., Naeff, H., Hannes Böttcher, H., 2012. Atlas of EU biomass potentials, Deliverable 3.3: Spatially detailed and quantified overview of EU biomass potential taking into account the main criteria determining biomass availability from different sources.

Ellen MacArthur Foundation, 2015. Towards a circular economy: business rationale for an accelerated transition. Ellen MacArthur Foundation.

European Commission, 2004. Commission Regulation (EC) No. 574/2004 of 23 February 2004 amending Annexes I and III to Regulation (EC) No. 2150/2002 of the European Parliament and of the Council on Waste Statistics (Text With EEA Relevance), Official J. L 90, 15–47.

European Commission, 2015. Proposal for a Directive of the European Parliament and of the Council Amending Directive 94/62/EC on Packaging and Packaging Waste, COM/2015/0596 final—2015/0276 (COD).

European Council, 1999. Council Directive 1999/31/EC of 26 April 1999 on the landfill of waste. Official J. L 182, 0001–0019.

European Parliament and the Council, 2008. Directive 2008/98/EC of the European Parliament and of the Council of 19 November 2008 on waste and repealing certain Directives (Text with EEA relevance). Official J. L 312, 3.

European Parliament and the Council, 2009. Regulation (EC) No. 1069/2009 of the European Parliament and of the Council of 21 October 2009 laying down health rules as regards animal by-products and derived products not intended for human consumption and repealing Regulation (EC) No. 1774/2002 (Animal by-products Regulation) Official J. L 300, 1–33.

Franckx, L., van Acoleyen, M., Hogg, D., Gibbs, A., Elliott, T., Sherrington, C., et al., 2009. Assessment of the options to improve the management of bio-waste in the European Union, Annex A: Baseline scenarios, Version A, 30-11-2009, Study contract No. 07.0307/2008/517621/ETU/G4, Arcadis and Eunomia for European Commission, DG Environment.

Franckx, L., van Acoleyen, M., Hogg, D., Gibbs, A., Elliott, T., Sherrington, C., et al., 2010. Assessment of the options to improve the management of bio-waste in the European Union, Final Report, Version C, 12-02-2010, Study contract No. 07.0307/2008/517621/ETU/G4, Arcadis and Eunomia for European Commission, DG Environment.

Jungmeier G., 2005. Cost Action E31, Management of recovered wood, Joanneum, training course, Hamburg.

LAP II, 2014. Landelijk afvalbeheerplan 2009–2021, Naar een materiaalketenbeleid, sector plan 36: hout. Dutch national waste management plan II, Sector plan 36: wood.

Leek N., 2010. Post-consumer wood, pp. 119–123. In: EUwood—Real Potential for Changes in Growth and Use of EU Forests. Methodology report. Hamburg/Germany, 165 pp.

Mantau, U., Steierer, F., Verkerk, H., Lindner, M., Antilla, P., Asikainen, A., et al., 2010. EUwood—Real Potential for Changes in Growth and Use of EU Forests. Final report. Hamburg/Germany, 160 pp.

PAS 111, 2012. PAS 111:2012 Specification for the requirements and test methods for processing waste wood. British Standards Institution ISBN 978-0-580-69643-5.

Probos, 2014. De markt van resthout en gebruikt hout in 2012, Bosberichten 2014-04 (in Dutch).

RecyclingholzV, 2012. 160. Verordnung des Bundesministers für Land- und Forstwirtschaft, Umwelt und Waserwirtschaft über das Recycling von Altholz in der Holzwerkstoffindustry. Austrian Ordinance for the recycling of wood waste in the wood materials industry.

Siemons, R., Vis, M., Van den Berg, D., I, M.C., Whiteley, M., Nikolaou, N., 2004. Bioenergy's Role in the EU Energy Market, a View of Developments Until 2020. Biomass Technology Group (BTG), Energy for Sustainable Development (ESD), Centre for Renewable Energy (CRES), Enschede, The Netherlands, p. 170. (plus Appendices).

Skovgaard, M., Moll, S., 2005. Outlook for Waste and Material Flows, Baseline and Alternative Scenarios. European Topic Centre on Waste and Material Flows, Copenhagen, Denmark.

Thrän, D., Weber, M., Scheuermann, A., Fröhlich, N., Zeddies, J., Henze, A., et al., 2006. Sustainable Strategies for Biomass Use in the European Context. Institut für Energetik und Umwelt (IE), Leipzig, Germany, p. 337. (plus Appendices).

UNECE (United Nations Economic Commission for Europe), FAO (Food and Agricultural Organization of the United Nations), 2011. The European Forest Sector Outlook Study II 2010–2030, Geneva.

Vis, M., 2016. Barriers to cascading use, presented at the 6th ISG Meeting of the Study on the Optimised Cascading Use of Wood, Brussels.

Vis, M., Dees, M., 2011. Biomass Resource Assessment Handbook, Harmonization of Biomass Resource Assessments, Best Practices and Methods Handbook. VDM Verlag Dr. Müller. ISBN: 978-3-639-29018-9.

Vis, M., Mantau, U., Allen, B., 2016. Study on the Optimised Cascading Use of Wood. Contract No. 394/PP/ENT/RCH/14/7689. Final report. Brussels, 337 pp. <http://bookshop.europa.eu/en/cascades-pbET0416305/>.

VLAREM, 1991. Besluit van de Vlaamse regering van 6 februari 1991 houdende vaststelling van het vlaams reglement betreffende de milieuvergunning. Flamish environmental permit ordinance.

Chapter 9

Assessing the Potentials for Nonfood Crops

Jacqueline Ramirez-Almeyda, Berien Elbersen, Andrea Monti, Igor Staritsky, Calliope Panoutsou, Efthymia Alexopoulou, Raymond Schrijver and Wolter Elbersen

9.1 INTRODUCTION

Dedicated perennial biomass crops still have a small area coverage in Europe. In the EU27 there were 5.5 million hectares used for dedicated biomass crops (for energy) and the dominant share (81%) was for oil crops (rape and sunflower) while only 1% was used for perennial lignocellulosic biomass crops (ETC/SIA, 2013; Elbersen et al., 2012 and Lewandowski et al., 2015) estimated the perennial area share in the same range at approximately 60,000 ha for EU-27. This confirms that perennials are not easily integrated into existing farming systems. There are several reasons why increased demand for lignocellulosic biomass will not necessarily lead to large production of perennials on good quality arable lands (Dees et al., 2017).

Firstly because these perennials cannot compete with food and feed crops unless low productive soils are involved where perennials may give higher yields and returns. Secondly, farmers are usually not willing to lose their flexibility by turning their land into long-term perennial plantations. After all they want to flexibly respond to market changes. Perennial plantations with a long lifetime of 15–20 years do not fit easily with this preference. On the other hand, experiences with certain perennial crops have shown that they have the capacity to deliver relatively high yields on lands of lower quality, which generally deliver minor noncompetitive yields for rotational food and feed crops (as was already shown in several EU research projects namely 4FCROPS, OPTIMA, OPTIMISC, WATBIO, SEEMLA, SRC plus, BFF, BioC4, MISCOMAR, and FIBRA, see Box 9.1 for information). The lower

Modeling and Optimization of Biomass Supply Chains.
DOI: http://dx.doi.org/10.1016/B978-0-12-812303-4.00009-4

BOX 9.1 Relevant EU Projects Evaluating Biomass Crops Suitability and Performance

4FCROPS project (www.cres.gr/4fcrops) completed in 2010 and delivered a lot of reports and publications on the most viable crops for nonfood biomass production (either for energy production or for bio-based materials) in every environmental zone in the EU, cost structure, and economic and environmental performance of the different crops. The project also made an estimate of the land availability for these crops at different time frames which could also provide further information on land identification.

Crops2Industry (http://www.crops2industry.eu/) was completed in 2012 and provided data on biomass crops that are best suited for selected industrial applications, namely oils, fibers, resins, pharmaceuticals, and other specialty products. Information was provided on yield and tolerance to abiotic and biotic conditions for different crops, while at the end the project provided output of an integrated technical, environmental, and economic assessment in the form of best nonfood crops for every location and end use in the EU.

OPTIMA (http://www.optimafp7.eu/) was an EU project completed in 2015 that aimed at identifying high-yielding perennial grasses for the Mediterranean area, within optimized production chains (for both energy and new plant derived bio-products). The focus was particularly on identifying and evaluating the best performing crops, genotypes, and farm management systems when grown on underutilized and/or abandoned marginal in Mediterranean environments.

FIBRA (http://www.fibrafp7.net/) project (2012–15) collected and analyzed valuable information on most suitable fiber crops and genotypes as sustainable sources of bio-based material for industrial crops covering the whole production chain and following the bio-refinery concept.

Water4Crops (http://www.water4crops.org/) can provide information on suitable crops and their performance of bio-treatment of wastewater in Europe.

OPTIMISC—Optimizing Miscanthus Biomass Production (https://optimisc.uni-hohenheim.de/). The overall objective of OPTIMISC was to optimize the *miscanthus* bioenergy and bio-products chain by trialing elite germplasm types over a range of sites across Europe; analyzing the key traits that currently limit the potential of miscanthus; identifying high-value bio-products; modeling the combined results to provide recommendations to policy makers, growers, and industry.

GrassMargins (http://www.grassmargins.com/). The project studies the potential for using different types of grass species under challenging climatic conditions (e.g., drought, salinity, flood, and cold) to develop high-yielding biofuel crops. The selection of material by partners in the project focuses on *miscanthus* and *giant reed* taxa.

quality lands are usually on the soils that first go out of conventional agricultural use. If a market indeed develops for lignocellulosic crops, it is therefore likely that the lands that are no longer used for conventional cropping with food and feed crops are partly used for production with crops like miscanthus, switchgrass, giant reed, and short rotation coppice (SRC) crops (Dees et al., 2017). In S2Biom we therefore aimed to obtain a better understanding of where unused lands will be in 2020 and 2030 (see Chapter 2: Existing Modeling Platforms for Biomass Supply in Europe: Inputs, Outputs, and Projection Potentials) and what crops are suited to be grown on these lands (focus of current chapter).

In this chapter an overview is provided of the main characteristics of perennial crops for which prospects exist to grow them in Europe to supply part of the increasing lignocellulosic biomass demands. From the published material available key phenological parameters were derived to simulate the growth of different perennial grasses and short rotation coppice (SRC) given the wide range in climatic and soil circumstances in Europe. How this was done is described in Section 9.1.1. In Section 9.2 it is discussed how the cost of production of these different cops was assessed taking account of the simulated yield levels for the whole of Europe. The chapter finalizes with a discussion of the main conclusions and the challenges, which aims to improve our knowledge based on dedicated biomass cropping in Europe, particularly in marginal lands.

9.1.1 Dedicated Biomass Cropping in Europe

Sustainability for nonfood biomass use, particularly for bioenergy and even more specifically for crop-based biomass resources, is an important issue in the policy and scientific debate for several years now. It is not surprising that there are many (global) studies that show that the shift from conventional energy production, with a negligible land demand, to low-carbon energy sources, including bioenergy, is a major driver of land use change and related greenhouse gas (GHG) emissions in the future (Valin et al., 2016; Laborde, 2011; Searchinger et al., 2008; Al-Riffai et al., 2010; E4tech, 2010; Edwards et al., 2010).

This is further exacerbated through the ambitions to shift toward a green circular economy in which more bio-based products will take the place of fossil-based products. The additional demand for biomass is therefore expected to increase even further as is discussed in Chapter 5, Modeling Demand for Bio-Based Sectors, and this demand cannot be satisfied only by residual and waste biomass sources. Hence it is likely that part of the demand for lignocellulosic biomass in the future will need to be addressed by dedicated cropping. This may increase the pressure on land that is already large and expected to increase even further because of a growing and changing global population, reaching ambitious bioenergy targets and bio-based economy ambitions, adapt to climate change while protecting the environment and the functioning of ecosystems (see also Chapter 2: Existing Modeling Platforms for Biomass Supply in Europe: Inputs, Outputs, and Projection Potentials).

Most critical issues in relation to the environmental sustainability of cropped biomass are (EEA, 2013; ETC/SIA, 2013):

1. land use change effects and related direct and indirect GHG emissions;
2. (direct and indirect) land use change (LUC and iLUC) and related biodiversity effects;
3. effects on water quality and quantity;
4. soil quality effects (soil organic carbon, soil erosion, nutrients).

Basically all these effects are determined by the combination of type of land and soil chosen, the type of crop used and related yielding capacity, and the management applied. The effects on biodiversity are determined both by direct uses of land and indirectly by the effects of the cropping activity on the overall environmental quality (soil, water, air).

The incremental use of land for agricultural production, whether a result of demand for biofuels, food, feed, or other nonfood applications, leads to an increase of GHG emissions and to loss of natural habitats with adverse effects on biodiversity. In spite of this iLUC related to demand for nonfood is more critically perceived than for food. The reason for this is that bioenergy and also biochemical production is stimulated principally to reduce GHG emissions (and related climate change) and thus decarbonize the energy and chemical sectors which still have great dependency on fossil feedstock (oil and natural gas). If through the iLUC effects the mitigation potential of bioenergy is diminished or even fully neutralized, one can question whether large stimulation of bioenergy and biochemical production only is advisable.

To avoid additional GHG emissions through iLUC effects, it is therefore preferred to grow dedicated biomass crops on lands that are marginal and abandoned for food and feed production as is also suggested in the low indirect impact biofuels methodology (Spöttle et al., 2013) and adopted by the roundtable on sustainable biomaterials. This avoids displacement effects and can enhance soil organic matter build up in marginal/abandoned lands (Monti and Zegada-Lizarazu, 2015; Cosentino et al., 2007).

From several field trials in a multitude of EU projects (Box 9.1) it has become clear so far that several perennial grasses and SRC crops are suitable to be grown in Europe and that their yielding capacities are promising also in lower productive lands.

The perennial grasses that have been investigated for Europe and for which there have also already been commercial plantations established include miscanthus (*Miscanthus* × giganteus Greef et Deuter), switchgrass (*Panicum virgatum* L.), giant reed (*Arundo donax* L.), reed canary grass (RCG) (*Phalaris arundinacea* L.), and cardoon (*Cynara cardunculus* L.). These have also been selected as the focus perennial crops in S2Biom. For the SRC crops, the most relevant for Europe are willow (*Salix* spp.), poplar (*Populus* spp.), and eucalyptus (*Eucalyptus* spp.).

Much relevant work on biomass crop selection and performance has been done and is being done in larger EU projects (see Box 9.1). A lot of valuable material is generated in these projects on identifying the best suitable perennial crops for bioclimatic and soil diversity in Europe in experimental fields and wider meta assessments by European crop experts. Several publications on the performance of these crops have also appeared in the last two decades (Lewandowski et al., 2015; Alexopoulou et al., 2015; Zegada et al., 2013; Cosentino et al., 2007; Triana et al., 2014; Fernandez, 2009; Christou et al., 2003; Hickman et al., 2010; Erickson et al., 2012; Sugiura, 2009; Guidi et al., 2008; Angelini et al., 2009; Price et al., 2004; Bassam, 2010).

9.1.2 Main Characteristics of Biomass Crops and Yield Potentials

In this section an overview is given of the main characteristics of perennial biomass grasses and short rotation types that can be grown in Europe based on recent publications and experiences gained in cropping experiments. The focus will be especially on the characteristics of these plant types in relation to factors that need to be taken into account when simulating their growth and identifying the soil and climatic circumstances where they can best be grown in Europe. This is then followed by a description of the crop growth simulation model that has been used to estimate yield levels for the whole of Europe for the different crops. The area cover 1480 NUTS3 (Nomenclature of Territorial Units for Statistics) code. In Section 9.3 the combined results of the yield simulation, yield suitability, and related cost levels are discussed.

9.1.3 Perennial Grasses

The perennial grass types presented in this chapter (Dees et al., 2017) are the ones mostly trialed and used in commercial plantations in Europe. They can be classified in C3 and C4 photosynthetic groups. C3 grasses are genetically most suited to temperate climates, while C4 grasses originate from tropical regions. Under cool weather conditions, C4 grasses produce less than C3 grasses, while in warmer climate zones, such as the Mediterranean zone, C4 grasses produce more. In addition, C4 grasses transpire less water per unit biomass than C3 grasses. An overview of perennial rhizomatous C3 and C4 grass characteristics is given in Table 9.1 and also in Tables 9.2 and 9.3 regarding soil and climate suitability.

TABLE 9.1 Biophysical Crop Data to Dedicated Cropping

Crop	Photosynthetic Pathway	Drought Tolerance	Water Requirement (mm)	Growing Tmax (°C)	Growing Tmin (°C)	GDD[a] (°C)	Kc[b]	WP[c] (g/L)	HI[d] (%)
Miscanthus	C4	High	>500	40	10	1700–2000	0.47–1.4	1.9–4.7	0.7
Switchgrass	C4	High	450–750	35	10	2060–2540	0.5–1.3	2.6–3.9	0.6
Giant reed	C3	High	380–650	35	5	1843–3000	0.5–1.7	2–6	0.7
RCG	C3	Medium	400–900	30	7	1800	1.24–1.46	1.5	0.6
Cardoon	C3	High	300–400	35	5	2425	0.5 –1.0	3.13	0.6
Willow	C3	Low	>620	30	0	2200	0.49–2.7	2.9–6.3	0.65
Poplar	C3	Medium	>600	30	0	2200	0.42–2.5	3.35–5.26	0.6
Eucalyptus	C3	Medium	>500	35	5	2400	0.50–1.8	1.64–4	0.65

[a] *GDD, growing degree days is the sum of daily temperatures from the start to end of the season to predict plant development rates.*
[b] *Kc, crop coefficient refers to a ratio between crop evapotranspiration and reference evapotranspiration.*
[c] *WP, water productivity refers to crop production in relation to total water consumed.*
[d] *HI, harvest index refers the ratio of yield to biomass. Some crops encounter more losses at harvest than others.*
Source: *Alexopoulou et al. 2010; Bassam, 2010; EEA, 2007; Zegada-Lizarazu et al. 2010; Lewandowski et al. 2003; Kretschmer et al. 2012; Fernandez et al. 2006.*

TABLE 9.2 Soil and Climate Limiting Factors per Perennial Crop Used to Generate Suitability Masks

	Classes	Miscanthus	Switchgrass	RCG	Giant Reed	Cardoon	Willow	Poplar	Eucalyptus
Slope (%)	<4%	4	4	4	4	4	4	4	4
	4–8	3	3	3	3	3	3	3	3
	8–15	2	2	2	2	2	2	2	2
	15–25	1	1	1	1	1	1	1	1
	>25%	0	0	0	0	0	0	0	0
Soil depth (cm)	Shallow (<40)	0	0	0	0	0	0	0	0
	Moderate (40–80)	1	1	1	1	1	1	1	1
	Deep (80–120)	2	2	3	2	2	2	2	2
	Very deep (>120)	4	4	4	4	4	4	4	4
Texture	Sand (coarse)	2	2	1	2	2	2	1	1
	Loam (medium–medium fine)	4	4	4	4	4	4	4	4
	Clay (fine)	2	2	2	3	2	2	2	2
	Heavy clay (very fine)	1	0	1	2	0	0	0	2
	Peat (no mineral texture)	0	0	0	0	0	0	0	0

(Continued)

TABLE 9.2 (Continued)

	Classes	Miscanthus	Switchgrass	RCG	Giant Reed	Cardoon	Willow	Poplar	Eucalyptus
Soil pH	0–4	0	0	0	0	0	0	0	0
	4–5	1	1	1	1	0	1	1	1
	5–6	3	2	3	2	2	2	2	2
	6–7	4	4	4	4	4	4	4	4
	7–8	2	2	2	2	3	3	3	3
Minimum and maximum temperature in GS (°C)	0–5	0	0	0	0	0	0	0	0
	5–8	1	1	1	0	0	1	1	0
	8–10	2	2	2	1	1	2	2	1
	10–20	4	4	4	3	3	4	4	3
	20–30	3	3	2	4	4	2	2	4
	Maximum temperature >30	1	1	0	2	2	0	0	2
Precipitation (mm)	0–300	0	0	0	0	0	0	0	0
	300–400	0	0	0	0	1	0	0	0
	400–500	0	1	0	1	2	0	0	0

	500–600	2	2	0	2	3	0	0	1
	600–800	3	3	2	3	3	2	2	2
	800–1000	3	4	3	4	4	3	3	3
Killing frost (°C)	>−20	0	0	2	0	0	2	2	0
	−20	0	2	2	0	0	2	2	0
	−10	2	2	2	0	0	2	2	0
	−5	2	2	3	2	2	3	3	2
	0	3	3	3	3	3	3	3	3

Note: The scoring on the different bio-physical factors is classified as follows: "0" unsuitable, "1" low suitability, "2" medium suitable, "3" suitable, "4" very suitable. Scores at level 0 or 1 are in the "marginality range" according to Van Oorschoven et al. (2012) and Terres et al. 2014.

Source: *Alexopoulou et al. 2010; Kretschmer et al. 2012; Terres et al. 2014; Zegada-Lizarazu et al. 2010; Perpiña Castillo, C. et al. 2015; Allen et al. 2014; Nsanganwimana et al. 2014; Fernandez, 2009; Alexopoulou et al. 2015; Hope and McElroy, A. 1990; Bassam, 2010; Angelini et al. 2009; Zub & Brancourt-Hulmel, 2010; Lewandowski et al. 2003; Fernandez et al. 2006.*

TABLE 9.3 Killing Frost Limiting Factors Applied Used to Generate Suitability Masks

Killing Frost	Miscanthus	Switchgrass	Giant Reed	RCG	Cardoon	Willow	Poplar	Eucalyptus
Winter (>5days)	−10	−20	0	−30	0	−30	−30	0
Spring (>2 days) GS[a]	0	−5	0	-	0	-	-	-

[a] *GS, growing season.*

For southern Europe the largest experience in field trials and commercial plantations exist with cardoon (*C. cardunculus* L.) and giant reed (*A. donax* L.), given good tolerance to dry conditions as they can survive well with a water availability in the growing season period of between 300 and 450 mm (Bassam, 2010) (Nsanganwimana et al., 2014) (see Table 9.1). For northern Europe the C3 RCG (*P. arundinacea*) is suitable, particularly because it can survive in very cold winters and it starts growing already at 5°C. The C4 miscanthus (*Miscanthus* × giganteus Greef et Deuter) and switchgrass (*P. virgatum* L.) can grow well over a wide European territory although like with all perennials they give highest yields when temperatures are moderate to high and there is sufficient rainfall well spread over the growing season (>500 mm). Both crop types can cope with temperatures below 0°C, but certain varieties of switchgrass can tolerate lower temperatures better than most miscanthus varieties do (see also Table 9.3). An important advantage of miscanthus is that it can produce high yields in a relatively short growing season (see growing degree day (GDD) in Table 9.1). Also RCG needs a relatively low number of GDD to go through all growth stages, although yields are not as high as can be reached with miscanthus which is related with the very high water efficiency for the latter (see Table 9.1).

9.1.4 Short Rotation Coppice

The SRC types selected are willow (*Salix* spp.), poplar (*Populus* spp.), and eucalyptus (*Eucalyptus* spp.) for which commercial plantations and trials for biomass production are commonly spread all over Europe. The trees used are managed specifically to be harvested every 3–5 years. The management practice takes advantage of the fact that trees make new growth from the stump or roots if cut down. In an SRC plantation when tree stems are cut down, new shoots will emerge which can be cut again after 3 years.

The main characteristics of the three SRC types have also been included in Table 9.1 and show that all types are in the C3 photosynthetic regime. Minimal water requirements for willow and poplar to survive are generally higher than for eucalyptus, and certainly all perennial grasses. The minimal GDDs are fairly long as compared to most perennials, but are more easily reached because they start growing at relatively lower temperature.

9.1.5 Crop Yield Simulation Approach

The biomass yield depends on very site-specific conditions such as soil, temperature, and water availability. A direct relation exits between biomass production and water consumed through transpiration. This section describes a simple model for an herbaceous yield estimation using the principal concepts of *AquaCrop* model from the FAO-56 (Steduto et al., 2009). It was used to simulate yields of the perennials for a water limited (no irrigation) and maximum yield situation. The major growth controlling climatic factor is the temperature-defined length of growing season, and within this growing season the most important factors are the solar radiation, temperature, evapotranspiration, and available water from soil moisture reserves and rainfall.

The maintenance of a green cover implies permanent vegetative growth, which requires that the grass and SRC plants form sufficient new shoots all the time, and that fully grown older shoots are removed, which calls for carefully designed management schemes, adapted to specific species and ecological conditions. However, in practice it is not possible to maintain a full season green cover, as the grass plants require an establishment period at the beginning of the growth cycle, or a recovery period after the winter, and, especially if the crop is allowed to flower, it has a senescence period after flowering, characterized by decay and yellowing of green leaves. Also the accumulated living biomass itself requires maintenance respiration, which causes internal competition for assimilates between growth of new biomass and maintenance of existing living biomass. The growth and greening cycle can also be influenced by cutting and regrowth schemes, which are very specific per species.

To simulate the yield of the biomass crops, the data on daily weather factors (MARS, 2014) are combined in the *AquaCrop* model (Steduto et al., 2012) with the phenological factors (presented in Table 9.1). These factors are needed for the estimation of the start date and length of the growing season, and the total water use and associated crop biomass production, e.g., total length growing season, minimum start day, GDD, minimum temperature above which the crop becomes active, water requirement, crop coefficient (Kc), water productivity (WP). Furthermore, they are needed to estimate the crop water transpiration during the growing season. The growing season is divided into four crop coefficient "Kc" phases (initial, crop development, midseason, late season) with specific crop water requirements for each Kc phase (Triana et al., 2014). Once the total evapotranspiration of a crop during the growing season has been calculated, the gross aboveground production (dry biomass) can be calculated with the WP expressed in gram dry biomass per liter water in a maximum and a water limited situation as explained in the following formulas:

9.1.5.1 Maximum Yield Potential (Y_{pot})

The productivity potential of biomass in response to water in Mg/ha d.m., where water is a no limit.

$$B_{pot} = \Sigma Tr^{*} WP \tag{9.1a}$$

$$Y_{pot} = B_{pot}{}^{*} HI \tag{9.1b}$$

9.1.5.2 *Yield in Water Limitation* (Y_{wl})

A *water limited yield* is simulated assuming that water availability is limited by the precipitation and related crop transpiration. Biomass productivity limited by water in the growing season in Mg/ha d.m. is used as a reduction factor (F_{wl}) that comes from the global water satisfaction index (GWSI) model, the factor depends on the photosynthetic system C3 or C4.

$$B_{wl} = \Sigma Tr^{*}F_{wl}{}^{*}WP \tag{9.2a}$$

$$Y_{wl} = B_{wl}{}^{*}HI \tag{9.2b}$$

where:

Y_{wl} and Y_{pot} are yield potential and yield under water limitation (Mg/ha d.m.)
B_{wl} and B_{pot} are biomass potential and biomass under water limitation (Mg/ha d.m.)
ΣTr is sum daily crop transpiration in the growing season (mm)
Kc is crop coefficient calculated for each growth stage
WP is the water productivity (kg/mm^3)
HI is the harvest index (%)
F_{wl} is the reduction factor to water limitation (%)

The AquaCrop model explained in the former simulates crop yield in response to available water, radiation, and temperature. It does, however, not take full account of extreme climatic and soil limiting factors which make certain regions of Europe completely unsuitable to cultivate certain crops. Therefore, an additional step had to be made in our analysis which was the elaboration of masks specific per crop that indicate where soil and/or climate limitations exist that make areas unsuitable. For the elaboration of these spatial masks a matrix was developed which matches with the classified soil and climate limitations with the specific crop requirements. The selection of parameters and classes builds on the Joint Research Centre (JRC) study (Terres et al., 2014) but has been refined using several references specified under Tables 9.2 and 9.3. In the tables the scoring per perennial crop is presented. The combination of factors in Table 9.2 scoring "low suitable" is completely masked out on a map for the specific crop. Killing frost factor was taken as a separate factor dominating over all other suitability factors in Table 9.2. For killing frost a distinction was made between winter frost (when the plant is dormant) and spring frost, when the growing season has started. Frost occurrence in this early growth stage can be particularly harmful for some crops (see Table 9.3) such as cardoon, giant reed, and eucalyptus limiting the area in Europe they can grow significantly as compared to switchgrass and also *miscanthus*. The latter crop is however not able to cope with too extreme winter colds as it limits strongly the survival rate and prevents enough regrowth in spring. This explains a slightly smaller area coverage for *miscanthus* as compared to switchgrass or willow.

The temperature range indicator in Table 9.1 shows the minimum and maximum temperatures a crop has to cope with in the growing season. A small difference in temperature such as in the first three classes in Table 9.3 is typical for the Boreal and Alpine north zones of Europe where growing seasons are very short and temperatures usually do not come far above 10°C. In these regions it is not really worthwhile to grow biomass crops as yields will remain very low and most of the perennials cannot reach their minimal GDDs to deliver good quality biomass. Furthermore, these regions also cope quite often with shallow or very peaty soils, which are other factors ruling out large areas in these northern European zones.

9.1.6 Crop Yield Levels and Management

The two yield estimates that have been made, one under full irrigation and one under purely rainfed conditions, assume optimum management conditions and should correspond with the highest yields observed in field experiments. In reality, the practical yields are often lower, due to suboptimum conditions such as nutrient shortage or incomplete plant cover.

Management of the crop is therefore an important factor determining the practical yield levels that can be reached. Simulated maximum yield levels therefore always need correction for management to translate them into practical yields. High inputs through management only make sense if this also results in high yields. High yields can only be reached if water and soil quality is not a limiting factor, or can be compensated enough through management (inputs). If the soil quality is poor, management needs to be tuned to lower inputs, as poor soil conditions, e.g., shallow soils, are challenging to improve, particularly from an economic perspective. Since it is the expectation that it is more sustainable to produce lignocellulosic crops on lower quality soils which have been left unused for food production, it is important to determine yields and cost in both high input systems as in medium and low input systems which enables to match the yield simulation results with all land–soil quality groups. The maximum and water limited yield simulation results were further converted into three types of yield–input–management levels: high, medium, and low input following the rules presented in Table 9.4. The three yield management level combinations are inputs to the activity-based costing (ABC) cost model (see Section 9.2).

TABLE 9.4 Assumptions Applied on Yield Maximum Potential (Y_{pot}) and Water Limited Yield (Y_{wl}) for a Further Translation in Three Attainable Biomass Yield Levels (ETC/SIA, 2013)

Input System Level	Irrigation	Explanation
High	Irrigation is applied	All crops could reach 90% of Y_{pot}
Medium	No irrigation is applied, with the exception of some water gift in the establishment phase for some dedicated crops if needed.	Attainable biomass yield equals the lowest value of the following two modeled yield levels: (90% of Y_{pot}) and (100% of Y_{wl}).
Low	No irrigation was applied	The yield level is dictated by water limited conditions (the lowest value of either 80% of Y_{wl}, or 50% of Y_{pot})

9.2 COST LEVELS FOR BIOMASS CROPS

For the calculation of cost of dedicated cropping biomass, an excel-based ABC model was developed to calculate the roadside cost per ton. This implies that the cost is only covered in this model up to the farm gate.

The ABC model covers the whole production process of alternative production routes that can be divided in logical organizational units, i.e., activities. The general purpose of this model is to provide minimum cost prices for the primary production of biomass feedstock at the roadside. ABC generates the costs of different components based on specific input and output associated with the choice of the means of production, varying with the local conditions and cost of inputs (e.g., labor, energy, fertilizers, lubricants, etc.). Since the production of most biomass is spread over several years, often long-term cycles in which costs are incurred continuously while harvest only takes place once in so many years, the net present values (NPV) of the future costs are calculated. In this way costs for production of annual and perennial crops, agricultural residuals, and forest biomass are made comparable (=all expressed in present Euros).

9.2.1 ABC Approach

In order to assess the cost of a dedicated crop per location in Europe, eight interrelated excel work sheets in the ABC model need to be filled. This enables calculation of dedicated biomass net present (NPV) cost per type of crop, in a 60-year cycle for every NUTS3 region in Europe for low, medium, and high input management systems.

The plantation lifetime per crop type assumed is as follows:

- 12 years for SRC willow, poplar, and eucalyptus and cardoon
- 15 years for perennial grasses.

In Fig. 9.1 an overview is given of the eight model modules involved in the calculation of the dedicated cropping cost:

1. *Crop input 1* module sheet gathers the potential yield levels of the selected crop and water use combination at regional level (NUTS3) from the crop yield simulation model for dedicated crops as discussed in the previous section.
2. *Crop input 2* module sheet gathers the condensed information of the selected crop covering information on inputs specific to the yield type, yield level, and input level selected in the case input sheet. It includes the following:
   a. Irrigation need m^3/ha (from the selection in crop inputs 1) specific per yield–management combination (so only applied in high input systems).
   b. Plant material input value in seed or rhizomes expressed in €/ha, based on expert knowledge and judgment for each crop. For distribution over management systems, we assume that these costs are at 100% for medium input systems, for low it is set to 90%, and for high it is set to 110%.
   c. Crop protection inputs for insecticides, fungicides, herbicides, and growth regulators in €/ha per treatment. For distribution over management systems, we assume that these costs are at 100% for medium input systems, for low it is set to 50%, and for high it is set to 150%.
   d. Fertilizers in €/ha. The value is calculated in the sheet using information on the nutrient content specific for the selected crop (N, P_2O_5, K_2O, MgO, S, CaO) which is also included in this sheet for every crop. These nutrient content levels have been collected in S2Biom in the database with biomass characteristics (see Chapter 3: Lignocellulosic Biomass Quality: Matching Characteristics With Biomass Conversion Requirements). Fertilization application rates are equal to this removal rate plus an extra 33% application for losses of nutrients. The fertilizer costs take account of the application level needs per crops and of national specific price levels for fertilizers. The national price levels are included in the "*national inputs*" data sheet.
3. In the *crop input 3* module, the different crop management activities are gathered according to the crop management selection in the case input. In order to make crops with different cropping intervals comparable through time, activities are set on or off per year over a 60-year cycle. With a total cycle of 60 years, complete cycles of 1-, 2-, 3-, 5-, 10-, 12-, 15-, 20-, or even 30- and 60-year intervals can be covered by the model.

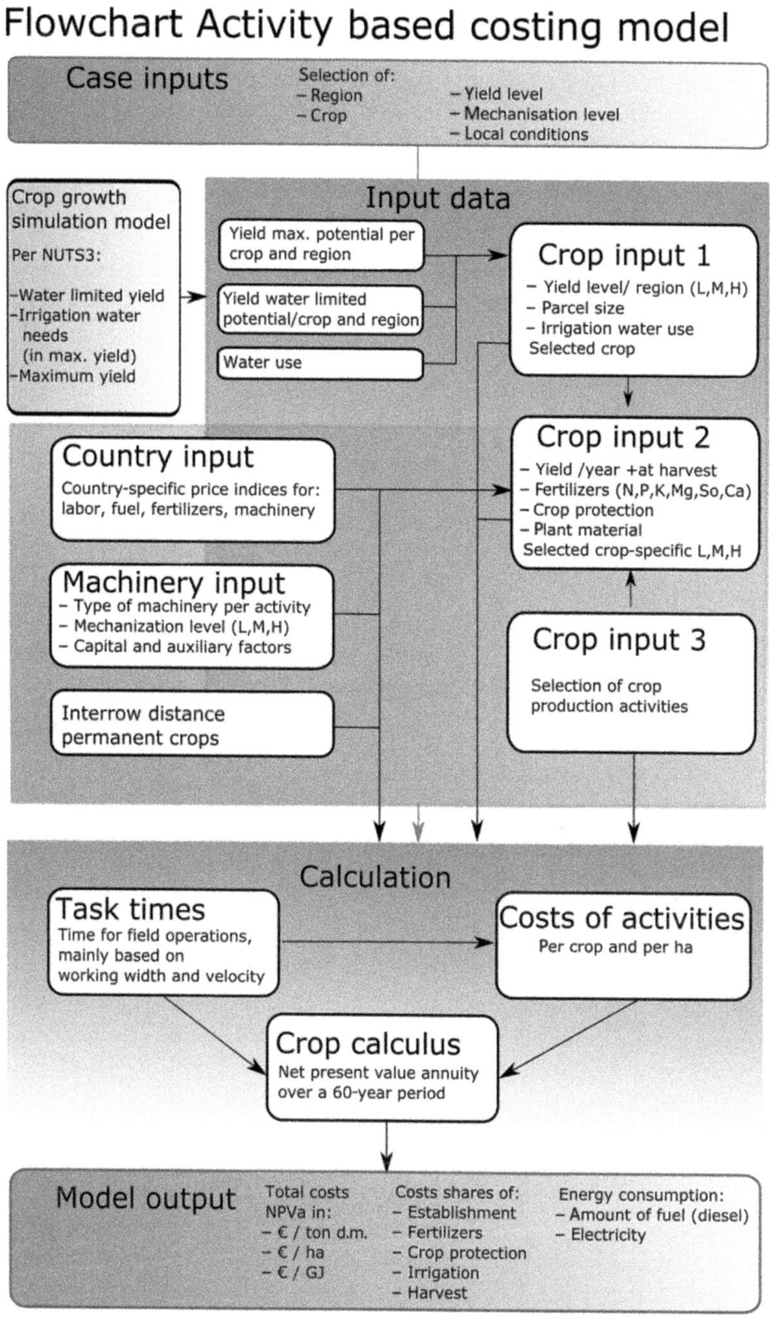

FIGURE 9.1 Overview of ABC cost calculation model for dedicated biomass crops.

4. In the "*country inputs*" module, detailed information concerning the prices of various resources needed as input for the production process of biomass are gathered specific per country. National level price data (e.g., VAT) included cover cost/prices for labor (skilled, unskilled, and average), fuel, electricity, fertilizers (N, P_2O_5, K_2O), machinery, crop protection, and land. Most of these data were gathered from statistical sources such as FADN (Farm Accountancy Data Network), Eurostat, and OECD. Most cost levels were gathered for the year 2012.
5. The "*machinery inputs*" module contains extensive information about different aspects of mechanized equipment involved in field operations. An average machine input mix was chosen associated with every activity but specified per management level. So larger machines are generally used in high input systems if these can be operated on the average field sizes encountered in the region under focus. The average field sizes as collected in the European land cover/use statistics database LUCAS (Eurostat, 2015) and gathered in the sheet *crop inputs 1* are determining the mechanization possibilities since the high level (and more expensive) equipment tend to find profitable application only on larger fields. The main source of information used to gather all financial and functional characteristics of the different machines is the German online database "KTBL-online" and WUR (KWIN). Low, medium, and high machinery input alternatives from this list have been chosen as typical instances.
6. In the "*task time activity*" module for a number of activities (field—operations), the amount of time needed to fulfill the operation is calculated with a model from De Lint et al. (1969). Per activity a calculation is made of the actual time consumed.
7. All modules described in the former then transfer the parameters to the *calculus* module which consists of two linked calculation sheets:
   a. The first is the "*activity cost calculus*" in which all activities cost are gathered from the different input sheets to calculate the total cost of all activities involved in terms of machinery, labor input, and energy input requirements.
   b. The time consumption per activity is gathered from the "*task time activity*" module as discussed above.
8. The final calculation worksheet which generates the final cost is the "*crop calculus*" module. In this final worksheet, all the costs associated with the selected crop in the specified region are expressed in a net present annuity, making each crop comparable through time. It adds up for the specific case input selection all activity cost, and all input cost per year, for the total lifetime of the plantation and repeats this until it reaches a total of 60 years. Cost are presented as NPV per annum and expressed in € per ton d.m. or per GJ. Net present values of all activities are calculated as follows:

$$\mathrm{NPV} = \mathrm{Fv}/(1 + i)n$$

where:

NPV = net present value
Fv = future value
i = the interest rate used for discounting (set to 4%)
n = number of years to discount

Then the NPV annuity is applied assuming that the sum of NPVs cover the annual capital payments attracted against the same interest rate (4%) as the discount rate used for calculating the NPVs.

$$NPVa = \Sigma NPV^{*}(1/((1-(1+i)-n)/i))$$

where:

NPVa = net present value annuity
ΣNPV = sum of NPVs
n = number of years
i = the interest rate (set to 4%)

9.3 RESULTS: COST–YIELD COMBINATIONS PER CROP AND MANAGEMENT SYSTEM

9.3.1 Final Yield and Cost Levels

The results of the assessments provide an overview of cost–yield combinations for the three management levels assumed (Figs. 9.2 and 9.3). At national level the results are presented in Annex 1. Note that the results in the tables do not take account of the soil and climatic limiting factors. An important difference occurs between the cost of medium and high yields.

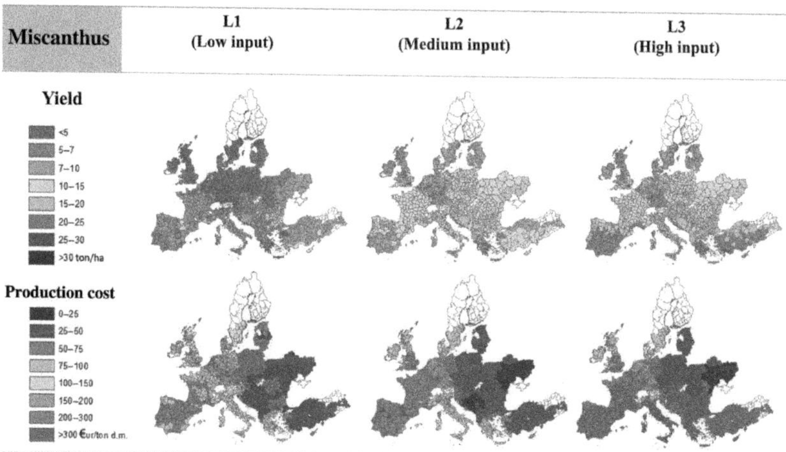

FIGURE 9.2 *Miscanthus* yield and cost levels for three different management systems (masks for regions unsuitable for these crops given soil and climatic limiting factors).

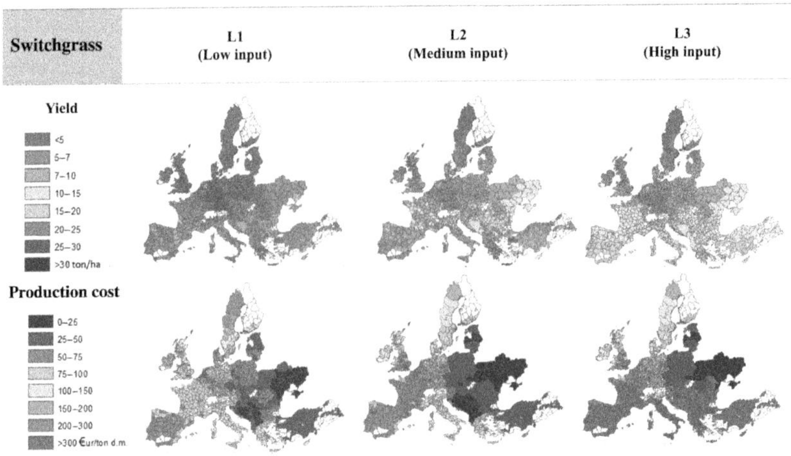

FIGURE 9.3 Switchgrass (*P. virgatum*) yield and cost level for three different management systems (masks for regions unsuitable for these crops given soil and climatic limiting factors).

This is mostly caused by the high irrigation cost that are needed to close the yield gap between a water limited yield and a maximum yield. This high cost difference is mostly seen in the southern and central European regions where precipitation is limited and irrigation can make a big difference in final yield. From an economic perspective, one can however conclude that applying irrigation is not rational as the additional yield does not compensate for the extra cost required. A small yield increase brought about by irrigation can already increase the cost by a factor 10 (Tables 9.A.1–9.A.3).

Cost of irrigation are high because it requires high investment cost in irrigation installations, fuel inputs for the pumping of water, and it needs to be done several times a year, while most of the other activities have a much lower frequency. The price of water itself was set at (0.01€/L) for all countries, which can be regarded as conservative as in several drought-prone countries prices of water can also be expected to increase and it may become common practice to charge farmers for irrigation water consumption, which until now is not common practice.

In the cost model irrigation costs were calculated applying four different technologies of which three were hose reel system with small to very large pumping capacities and a solar drip irrigation (20 ha, 75 kW electric pump). The first hose reel systems are much cheaper leading to low machinery investment cost, but require a lot of energy for the pumping of all irrigation water. The solar drip irrigation installation has very high investment cost, but does not need any fuel input. So the energy costs are zero, but the machinery costs are extremely high. The costs of irrigation become particularly high because irrigation application needs to be done several times a

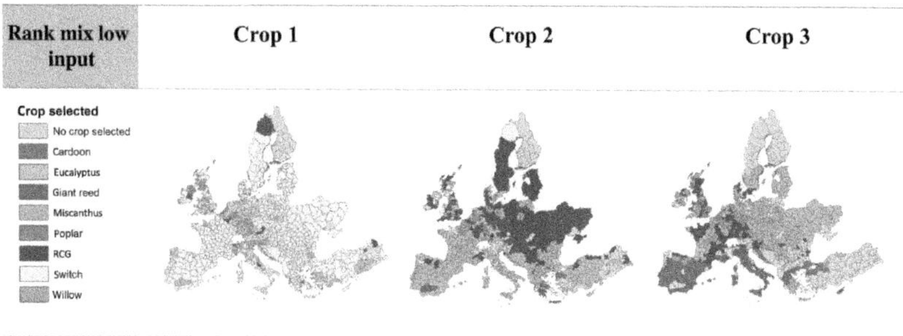

FIGURE 9.4 Top three of cheapest crop mixes for low input systems.

year, while most other activities take place in a perennial plantation once a year or less (e.g., sowing/planting, weeding, fertilization, and harvesting).

All options for calculating irrigation input result in very high cost levels. They confirm that irrigation in dedicated perennial cropping is just not an option, not even where it can lead to the doubling of the yields as the cost will increase by 20 times at least.

9.3.1.1 Best Resulting Crop Mix per Region

By combining the yield levels, the cost level, and the suitability masks per crop type, it could be determined per region what the best crop–management mix would be. In Fig. 9.4 the ranking of the top three crop–management combinations generating the lowest average NPV cost (€/ton d.m.) is presented in case of low input management systems which link to lower quality often characterized as marginal lands.

For the low input systems, perennials grasse are often the first choice, particularly for switchgrass and often as second miscanthus in the central and south and RCG in the north and east of Europe. Giant reed and cardoon are most often selected as a third, particularly in the southern regions. As to the SRC types they become selected at much lower frequency in the top three of low input crops. The one most selected is *Eucalyptus* in the southern regions and willow in the Northwest. In the central and eastern European countries, the picture is a mix of switchgrass, RCG, and *miscanthus*, while no SRC gets selected.

In the medium and high management systems, the ranking of crops does not become very different (Fig. 9.5) with perennials like switchgrass, miscanthus, giant reed in the medium management category mostly dominating in the first selection. The United Kingdom and Ireland are different in that there is a lot of willow-medium selected as first crop. In the second and third choice, there is an overall increase in RCG and miscanthus in the east and north of Europe. Toward the southern regions, there is an increase of giant reed and miscanthus as a second and third crop. In most cases the choices

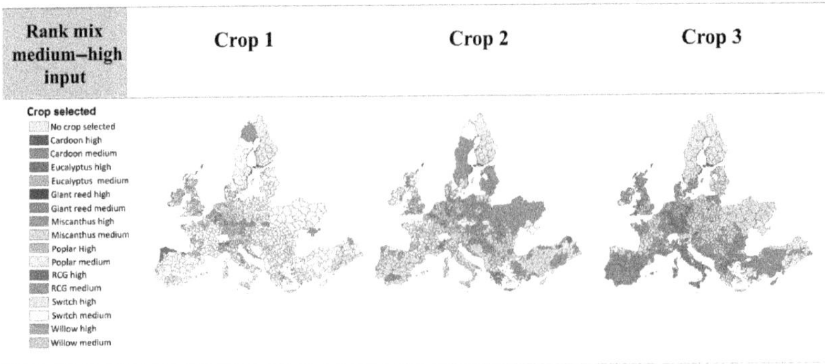

FIGURE 9.5 Top three of cheapest crop–management mixes for medium and high input systems.

are all in the medium management class, but sometimes in the regions of the western Balkans and Turkey there is also some high input miscanthus systems chosen as second or third option.

9.4 CONCLUSIONS: LIMITATIONS AND FURTHER RESEARCH CHALLENGES

Although much work has already been done on studying yielding capacity and soil and management of perennial biomass crops in Europe in field trials, a full extrapolation of the yielding capacity of these different crops for the whole of Europe has not been made before. In this respect the work presented here is novel. The results can give guidance to the choice of crop management combinations in different regions. A very important conclusion that can be drawn clearly is that irrigation is not likely to be economically feasible in most perennial crops and therefore likely to be less of an environmental threat in case of drought prone regions with large unused lands potential. It implies, however, that if dedicated cropping is to take place in southern Europe, there is an evident to find perennial crops and varieties that are strongly drought-tolerant and water use efficient.

Although the work performed delivers already some interesting observations, the results of the yield and cost level estimations for the different dedicated crops can certainly be further improved in methodology and data inputs in the future:

1. The experience with growing biomass crops in Europe is still limited and crop simulation input data were mostly based on field trial information from a limited number of locations. The yield simulation work done in this project was based on data available, but much more experience and field trailing is necessary to obtain more reliable crop yield response understanding, particularly in more marginal circumstances in Europe. Estimates of cost and yields for dedicated biomass crops should be based

as much as possible on larger scale field experiences in the relevant bioclimatic circumstances. The more field experience there is, the more reliable yield estimates can be made for the whole European territory using crop yield simulation models.

2. In this study no attention was paid to performance differences for different varieties of the same crop. For the crop simulation work crop phenological parameters and ecophysiological were taken that refer to an average crop performance not taking account of all variety characteristics that may exist. The simple assumption was made that the crop used in every region was chosen according to best performing varieties of the crop. So, for example, for southern Finland a switchgrass variety was chosen that proved to be best surviving in cold winters and short growing season.
3. Since there is still relatively limited experience in Europe in producing these crops, the costing of growing these crops is also rather immature. The experiences with specific field activities and also the machinery available are still limited and in development. Exact knowledge on management of the dedicated cropping systems can be improved when field experience is further incorporated in the cost calculation.
4. Currently limited costs are assumed for the preparation of the field before establishing the crop, while this could be very challenging in lands that have gone out of production for a longer period of time and/or are marginal or low productive. This aspect certainly needs further investigation.

In addition to the limitations in data and experience, it is also very challenging to make reliable predictions on the mobilization of dedicated crops on unused lands. Currently lignocellulosic and woody crop production areas are very small. Whether the potential from these crops will really be mobilized depends on many factors which were not all be taken into account in the assessment of the potentials for dedicated crops. Some of the main factors determining future mobilization are summarized below:

1. Access rights to lands.
2. Uncertainty about establishment cost, particularly on the least productive lands.
3. Uncertainty about economic returns as experience with growing these novel biomass crops is very limited and mostly at the stage of field trials. Some European-wide scattered commercial experiences exist which so far have not been proven to be very successful (e.g., RCG production in Finland). Uncertainty about economic returns is influencing the interest of land owners and/or investors.
4. Loss of flexibility by the farmer to decide on his/her cultivation plan as the plantations usually have a lifetime of between 10 and 20 years.
5. Unclear arrangements regarding Common Agricultural Policy (CAP) payment rights in certain EU countries when agricultural land is planted with SRC crops for a longer period of time.
6. Opportunity to set up optimal logistical biomass delivery chains making collection cost effective requires cooperation between different factors in

the supply chain not only several biomass providers (farmers), but also the conversion industries.

7. Ensure minimal feedstock delivery and security of supply which can only be realized if several farmers together invest in dedicated biomass cropping activities in one region.

Most of these factors were not taken into consideration in our study because they are difficult to predict.

9.5 LIST OF ABBREVIATIONS

ABC activity-based costing
CAP Common Agricultural Policy
EU European Union
ET_0 reference evapotranspiration
FADN Farm Accountancy Data Network
GHG greenhouse gas
GWSI global water satisfaction index
GDDs growing degree days
iLUC indirect land use change
JRC Joint Research Centre
MARS monitoring of agriculture with remote sensing
NUTS Nomenclature of Territorial Units for Statistics (of the European Union)
NPV net present value
LUC direct land use change
RCG reed canary grass
SRC short rotation coppice

REFERENCES

Alexopoulou, E., Christou, M., Eleftheriadis, I., 2010. Role of 4F cropping in determining future biomass potentials, including sustainability and policy related issues. A publication as part of the Biomass Futures project (IEE 08 653 SI2. 529 241) funded by the European Union's Intelligent Energy Programme, pp. 1–40.

Alexopoulou, E., Zanetti, F., Scordia, D., Zegada-Lizarazu, W., Christou, M., Testa, G., et al., 2015. Long-term yields of switchgrass, giant reed, and miscanthus in the Mediterranean Basin. BioEnergy Res. 8 (4), 1492–1499.

Allen, B., Kretschmer, B., Baldock, D., Menadue, H., Nanni, S., Tucker, G., 2014. Space for energy crops—assessing the potential contribution to Europe's energy future. Report produced for BirdLife Europe, European Environmental Bureau and Transport & Environment, IEEP, London.

Al-Riffai, P., Dimaranan, B., Laborde, D., 2010. Global Trade and Environmental Impact Study of the EU Biofuels Mandate. IFPRI.

Angelini, L.G., Ceccarini, L., Nassi o Di Nasso, N., Bonari, E., 2009. Comparison of *Arundo donax* L. and *Miscanthus x giganteus* in a long-term field experiment in Central Italy: analysis of productive characteristics and energy balance. Biomass Bioenergy 33 (4), 635–643.

Bassam, E., 2010. Handbook of Bioenergy Crops: A Complete Reference to Species, Development and Applications. Earthscan, London, UK, pp. 1–545.

Christou, M., Mardikis, M., Alexopoulou, E., Cosentino, S.L., Copani, V., & Sanzone, E. (2003). Environmental studies on *Arundo Donax.* 8th In: International Conference on Environmental Science and Technology Lemnos Island, Greece, 8 – 10 September 2003, (September), pp. 102–110.

Cosentino, S.L., Patanè, C., Sanzone, E., Copani, V., Foti, S., 2007. Effects of soil water content and nitrogen supply on the productivity of *Miscanthus* × *giganteus* Greef et Deu. in a Mediterranean environment. Indus. Crops Prod. 25 (1), 75–88.

Dees, M., Elbersen, B., Fitzgerald, J., Vis, M., Anttila, P., Forsell, N., et al., 2017. A spatial data base on sustainable biomass cost–supply of lignocellulosic biomass in Europe—methods & data sources. Project Report. S2Biom—a project funded under the European Union 7th Framework Programme for Research. Grant Agreement no. 608622. Chair of Remote Sensing and Landscape Information Systems, Institute of Forest Sciences, University of Freiburg, 174 pp.

De Lint, M.M. (1969) Uitgangspunten en berekeningsmethoden voor de taaktijdopbouw in het *nieuwe "Taaktijdenboek". Wageningen, Instituut voor Landbouwtechniek en Rationalisatie.*

E4tech, 2010. A causal descriptive approach to modelling the GHG emissions associated with the indirect land use impacts of biofuels. Final Report, E4tech, London. EEA (2007). Estimating the environmentally compatible bioenergy potential from agriculture. European Environment Agency (EEA). Technical report No 12/2007. p. 134.

EEA, 2007. Estimating the environmentally compatible bioenergy potential from agriculture. European Environment Agency (EEA), p. 134. Technical report No 12/2007.

Edwards, R., Mulligan, D. & Marelli, L., 2010. Indirect land use change from increased biofuels demand. Available at: <https://www.energy.eu/publications/Indirect-Land-Use-Change-from-increased-Biofuels-Demand.pdf>.

Elbersen, B., Staritsky, I., Hengeveld, G., Schelhaas, M.J., Naeff, H., Böttcher, H., 2012. Atlas of EU biomass potentials. Deliverable 3.3: Spatially detailed and quantified overview of EU biomass potential taking into account the main criteria determining biomass availability from different sources. Biomass Futures.

Erickson, J.E., Soikaew, A., Sollenberger, L.E., Bennett, J.M.,, 2012. Water use and water-use efficiency of three perennial bioenergy grass crops in Florida. Agriculture 2 (4), 325–338.

ETC/SIA, 2013. Review of the EU bioenergy potential from a resource efficiency perspective. Background Report to EEA study, Alterra, Wageningen.

Eurostat (2015). LUCAS data 2015 at: <http://ec.europa.eu/eurostat/web/lucas/data/primary-data/2015>.

Fernandez, J., 2009. El Cultivo de Cardo (*Cynara cardunculus* L.) para la produccion de biomas-sa. Hojas Divulgativas, n. 2130 HD1–44.

Fernandez, J., Curt, M.D., Aguado, P.L., 2006. Industrial applications of *Cynara cardunculus* L. for energy and other uses. Indus. Crops Prod. 24 (3), 222–229.

Guidi, W., Piccioni, E., Bonari, E., 2008. Evapotranspiration and crop coefficient of poplar and willow short-rotation coppice used as vegetation filter. Bioresource Technol. 99 (11), 4832–4840.

Hickman, G.C., Vanloocke, A., Dohleman, F.G., Bernacchi, C.J., 2010. A comparison of canopy evapotranspiration for maize and two perennial grasses identified as potential bioenergy crops. GCB Bioenergy 2 (4), 157–168.

Hope, H., McELROY, A., 1990. Low-temperature tolerance of switchgrass (*Panicum virgatum* L.). Can. J. Plant Sci. 70, 1091–1096.

Kretschmer, B., Alexopoulou, E., Panoutsou, C., 2012. Mapping the biomass crops options in 2020 and 2030 in EU27. Biomass Futures Project. CRES (Centre for Renewable Energy Sources). D6.4–WP6, July 2011. Retrieved from: <www.cres.gr/4fcrops>.

Laborde, D., 2011. Assessing the land use change consequences of European biofuel licies. Final report ATLASS consortium. Contract no. TRADE/07/A2. IFPRI, Washington DC, <http://trade.ec.europa.eu/doclib/docs/2011/october/tradoc_148289.pdf> (accessed 01.07.12.).

Lewandowski, I., Scurlock, J.M.O., Lindvall, E., Christou, M., 2003. The development and current status of perennial rhizomatous grasses as energy crops in the US and Europe. Biomass Bioenergy 25 (4), 335–361.

Lewandowski, I., Clifton-Brown, J., Murphy-Bokern, D., 2015. Perennial biomass Crops for a Resource-Constrained world. Biomass 2015 Summary. Stuttgart-Hohenheim, Germany, pp. 1–31. Available from: <www.biomass2015.eu>.

MARS, 2014. Gridded Agro-Meteorological Data in Europe, European Commission. Meteorological data available. MARS-AGRI4CAST Resources Portal. <http://agri4cast.jrc.ec.europa.eu/DataPortal/Index.aspx>.

Monti, A., Zegada-Lizarazu, W., 2015. Sixteen-year biomass yield and soil carbon storage of giant reed (*Arundo donax* L.) grown under variable nitrogen fertilization rates. Bioenergy Res. 9, 248–256.

Nsanganwimana, F., Pourrut, B., Mench, M., Douay, F., 2014. Suitability of *Miscanthus* species for managing inorganic and organic contaminated land and restoring ecosystem services. A review. J. Environ. Manage. 143, 123–134.

Perpiña Castillo, C., Lavalle, Carlo, Baranzelli, Claudia, Mubareka, Sarah, 2015. Modelling the spatial allocation of second-generation feedstock (lignocellulosic crops) in Europe. Int. J. Geograph. Inform. Sci. 29 (10), 1807–1825.

Price, L., Bullard, M., Lyons, H., Anthony, S., Nixon, P., 2004. Identifying the yield potential of *Miscanthus x giganteus*: an assessment of the spatial and temporal variability of *M. x giganteus* biomass productivity across England and Wales. Biomass Bioenergy 26 (1), 3–13.

Searchinger, T., Heimlich, R., Houghton, R.A., Dong, F., Elobeid, A., Fabiosa, J., et al., 2008. Use of U.S. croplands for biofuels increases greenhouse gases through emissions from land-use change. Science 319 (5867), 1238–1240.

Smit, H.J., Metzger, M.J., Ewert, F., 2008. Spatial distribution of grassland productivity and land use in Europe. Agric. Syst. 98, 208–219.

Spöttle, M., Alberici, S., Toop, G., Peters, D., Gamba, L., Ping, S., et al., 2013. Low ILUC potential of wastes and residues for biofuels. Straw, forestry residues, UCO, corn cobs. Project number: BIEDE13386/BIENL12798.

Steduto, P., Hsiao, T.C., Raes, D., Fereres, E., 2009. AquaCrop—the FAO crop model to simulate yield response to water: I. Concepts and underlying principles. Agron. J. 101, 426–437.

Steduto, P., Raes, D., Hsiao TCFE, 2012. Crop Response to Water. Irrigation and Drainage Paper. FAO, Rome, p. 66.

Sugiura, A., 2009. Waterrenew: Wastewater Polishing Using Renewable Energy Crops (Ph.D. dissertation). Cranfield University, School of Applied Sciences.

Terres, J.M., Hagyo, A., Wania, A. (Eds.), Confalonieri, R., Jones, B. Van Diepen, K., Van Orshoven, J., 2014. Scientific contribution on combining biophysical criteria underpinning the delineation of agricultural areas affected by specific constraints. Methodology and factsheets for plausible criteria combinations. JRC Science and Policy Report EUR 26940 EN, pp. 1–81.

Triana, F., Nassi o Di Nasso, N., Ragaglini, G., Roncucci, N., Bonari, E., 2014. Evapotranspiration, crop coefficient and water use efficiency of giant reed (*Arundo donax* L.) and miscanthus (*Miscanthus* × *giganteus* Greef et Deu.) in a Mediterranean environment. GCB Bioenergy811–819.

Valin H., Peters D., van den Berg M., Frank S., Havlik P., Forsell N., et al., 2016. The land use change impact of biofuels consumed in the EU, Quantification of area and greenhouse gas impacts. Report prepared by Ecofys/IIASA/E4tech for DG ENER under contract BIENL13120.

Van Oorschoven, J., Terres, J., Tóth, T. (Eds.), 2012. Updated common bio-physical criteria to define natural constraints for agriculture in Europe. Definition and scientific justification for the common biophysical criteria. JRC Report EUR 25203 EN, pp. 1–72.

Zegada-Lizarazu, W., Elbersen, W., Cosentino, S.L., Zatta, A., Alexopoulou, E., Monti, A., 2010. Agronomic aspects of future energy crops in Europe. Biofuels Bioprod. Bioref 4 (674–691), 14.

Zegada-Lizarazu, W., Parrish, D., Berti, M., Monti, A., 2013. Dedicated crops for advanced biofuels: consistent and diverging agronomic points of view between the USA and the EU-27. Biofuels Bioprod. Biorefin.

Zub, H.W., Brancourt-Hulmel, M., 2010. Review article agronomic and physiological performances of different species of *Miscanthus*, a major energy crop. A review. Agron. Sustain. Dev. 30, 201–214.

FURTHER READING

Elbersen, W., Poppens, R., Bakker, R., 2013. Switchgrass (*Panicum virgatum* L.) A perennial biomass grass for efficient production of feedstock for the biobased economy. NL Agency 28.

Elbersen, W., Bakker, R., Elbersen, B.S., 2005. A simple method to estimate practical field yields of biomass grasses in Europe. Paper presented as poster at the 14th European Biomass Conference, Paris.

Fazio, S., Barbanti, L.,, 2014. Energy and economic assessments of bio-energy systems based on annual and perennial crops for temperate and tropical areas. Renew. Energ. 69, 233–241.

Feigenbaum, S., Bielorai, H., Erner, Y., Dasberg, S., 1987. The fate of 15N labelled nitrogen applied to mature citrus trees. Plant. Soil 97, 179–187.

Fiorese, G., Guariso, G., 2010. A GIS-based approach to evaluate biomass potential from energy crops at regional scale. Environ. Modell. Softw. 25 (6), 702–711.

Kahle, P., Beuch, S., Boelcke, B., Leinweber, P., Schulten, H.R., 2001. Cropping of Miscanthus in Central Europe: biomass production and influence on nutrients and soil organic matter. Eur. J. Agron. 15 (3), 171–184.

Kandel, T.P., Hastings, A., Jorgensen, U., Olesen, J.E., 2016. Simulation of biomass yield of regular and chilling tolerant Miscanthus cultivars and reed canary grass in different climates of Europe. Indus. Crops Prod. 86, 329–333.

Katerji, N., Mastrorilli, M., Rana, G., 2008. Water use efficiency of crops cultivated in the Mediterranean region: review and analysis. Eur. J. Agron. 28 (4), 493–507.

Monti, A., Zanetti, F., Scordia, D., Testa, G., Cosentino, S.L., 2015. What to harvest when? Autumn, winter, annual and biennial harvesting of giant reed, miscanthus and switchgrass in northern and southern Mediterranean area. Indus. Crops Prod. 75, 129–134.

ANNEX 1 OVERVIEW OF YIELD-COST RESULTS

TABLE 9.A.1 Average Yield (ton d.m./ha/year) and Roadside Cost (€/ton d.m.) Low Management Level for EU28

	Miscanthus		Switchgrass		Giant Reed		RCG		Cardoon		Willow		Poplar		Eucalyptus	
Country	Yield	Cost	Yield	Cost	Yield	Cost	Yield	Cost	Yield	Cost	Yield	Cost	Yield	Cost	Yield	Cost
AL	6.9	38	5.1	27	7.8	46	4.4	30	4.8	40	6.7	35	6.2	37	5.4	41
AT	6.7	95	5.0	106	7.0	123	4.2	117	4.4	164	6.7	75	5.6	89	4.9	103
BA	10.0	28	7.4	22	10.1	39	5.6	25	6.3	34	8.4	29	8.1	30	7.0	32
BE	6.6	106	5.0	119	6.8	131	4.2	132	4.3	174	6.4	101	5.6	115	4.7	133
BG	9.9	53	7.4	54	10.8	62	5.1	75	6.6	72	7.6	58	7.7	56	7.5	54
CY	11.4	74	8.9	83	15.0	76	5.3	123	9.3	95	8.1	86	8.2	84	9.5	71
CZ	6.5	56	4.9	51	7.1	69	4.3	55	4.5	75	6.8	48	5.5	58	5.0	62
DE	6.5	127	4.8	148	6.9	153	4.2	166	4.3	205	6.7	103	5.5	120	4.9	131
DK	5.7	180	4.3	213	6.0	215	3.6	243	3.8	285	6.0	153	4.7	191	4.2	208
EE	4.7	59	3.5	44	4.9	75	3.0	50	3.2	73	4.8	54	4.0	63	3.5	71
EL	6.2	126	4.8	140	8.4	115	3.2	206	5.1	160	4.9	130	5.0	123	5.7	107
ES	8.0	70	6.1	72	9.4	85	4.2	96	5.8	102	6.3	74	6.3	72	6.4	67
FI	4.0	159	3.0	174	4.1	203	2.6	195	2.6	262	4.3	125	3.4	150	2.9	173
FR	8.0	81	6.0	88	8.5	105	4.8	106	5.2	133	7.2	75	6.6	80	5.9	85
HR	9.2	34	6.9	27	10.1	45	4.8	34	6.2	44	7.2	37	7.3	36	7.0	36
HU	8.9	50	6.6	48	9.7	64	4.9	60	6.0	67	7.4	50	7.2	50	6.8	50
IE	5.0	150	3.8	168	5.0	188	3.5	176	3.2	237	5.3	129	4.6	146	3.5	183
IT	8.5	111	6.5	129	10.0	129	4.3	172	6.1	165	6.6	119	6.6	117	6.9	108

KS	7.8	34	5.8	25	8.4	43	5.1	26	5.2	38	8.4	29	6.3	37	5.8	38
LT	5.5	60	4.1	50	5.9	77	3.4	62	3.7	76	5.7	51	4.5	61	4.1	62
LU	6.3	103	4.7	112	7.0	126	4.2	122	4.3	160	6.7	88	5.6	103	4.8	113
LV	5.3	50	3.9	35	5.5	66	3.2	40	3.5	60	5.3	47	4.3	56	3.9	60
MD	8.8	30	6.6	22	9.6	39	4.8	26	6.0	33	7.5	31	7.0	33	6.7	32
ME	9.0	31	6.7	23	9.8	40	5.9	24	6.1	35	8.9	27	7.8	31	6.9	33
MK	8.5	31	6.3	22	9.4	38	4.8	25	5.8	32	7.2	32	7.1	32	6.6	33
MT	10.2	178	7.9	224	13.3	178	4.7	380	8.2	262	7.1	187	7.3	180	8.5	151
NL	6.2	303	4.6	384	6.5	326	4.0	438	4.1	486	6.3	263	5.3	307	4.6	352
PL	6.2	62	4.6	56	6.8	72	4.0	63	4.3	78	6.5	54	5.2	65	4.8	69
PT	7.9	60	6.0	58	9.9	71	4.1	79	6.1	80	6.1	62	6.3	59	6.9	50
RO	8.4	51	6.3	48	9.1	63	4.7	59	5.7	65	7.3	47	6.8	49	6.3	49
RS	9.1	30	6.8	22	9.8	39	4.9	27	6.1	35	7.5	32	7.3	32	6.8	33
SE	4.4	150	3.3	163	4.6	193	2.9	177	2.9	247	4.8	122	3.8	151	3.2	178
SI	7.7	96	5.7	111	8.1	115	4.7	144	5.1	162	7.2	81	6.4	88	5.7	96
SK	7.6	43	5.7	36	8.3	55	4.7	39	5.2	55	7.3	39	6.3	45	5.8	48
TR	7.9	41	6.0	33	9.1	51	4.4	43	5.6	50	6.7	40	6.4	42	6.2	40
UA	7.6	35	5.6	24	8.1	44	4.5	27	5.1	37	7.1	33	6.2	37	5.7	39
UK	5.7	112	4.2	121	5.9	141	3.9	135	3.7	179	6.0	95	5.0	112	4.1	132

TABLE 9.A.2 Average Yield (ton d.m./ha/year) and Roadside Cost (€/ton d.m.) Medium Management Level for EU28

	Miscanthus		Switchgrass		Giant Reed		RCG		Cardoon		Willow		Poplar		Eucalyptus	
Country	Yield	Cost	Yield	Cost	Yield	Cost	Yield	Cost	Yield	Cost	Yield	Cost	Yield	Cost	Yield	Cost
AL	8.6	35	6.4	24	9.7	41	5.5	26	6.0	35	8.4	32	7.8	34	6.7	38
AT	12.0	58	8.9	64	12.4	77	7.4	70	7.8	100	11.8	46	9.9	55	8.6	62
BA	15.5	22	11.5	18	15.5	31	8.6	20	9.6	27	12.9	22	12.5	23	10.8	24
BE	12.0	63	8.9	70	12.2	80	7.6	76	7.7	103	11.6	60	10.0	68	8.5	78
BG	14.1	43	10.6	45	15.4	52	7.2	61	9.4	59	10.9	46	11.0	45	10.7	43
CY	17.7	53	13.9	60	23.1	58	8.2	84	14.3	69	12.5	60	12.7	59	14.7	49
CZ	11.8	36	8.8	33	12.6	46	7.7	34	7.9	49	12.1	31	9.8	37	8.8	39
DE	11.6	81	8.7	95	12.0	104	7.3	108	7.6	134	11.6	66	9.6	76	8.4	81
DK	10.2	113	7.7	133	10.7	140	6.4	152	6.8	179	10.8	93	8.5	115	7.5	122
EE	8.5	37	6.4	28	8.6	49	5.3	33	5.5	48	8.3	35	6.9	41	6.0	46
EL	7.8	107	6.0	116	10.6	97	4.0	168	6.4	132	6.1	111	6.3	105	7.1	90
ES	10.1	61	7.7	62	12.0	74	5.4	81	7.4	87	8.1	64	8.1	62	8.1	57
FI	7.2	100	5.4	111	7.4	131	4.6	123	4.7	164	7.6	78	6.1	93	5.2	105
FR	12.7	60	9.5	65	13.0	79	7.4	80	8.1	99	11.2	55	10.2	58	9.1	60
HR	13.9	26	10.5	21	15.3	35	7.3	25	9.4	34	11.0	28	11.1	28	10.7	27
HU	13.4	40	10.0	40	14.4	53	7.3	48	8.9	55	11.0	39	10.7	39	10.0	38
IE	9.0	95	6.8	106	9.0	123	6.3	111	5.8	149	9.6	79	8.3	89	6.3	109
IT	11.8	89	9.0	104	13.7	106	6.0	134	8.4	131	9.1	94	9.1	92	9.4	83

KS	12.1	26	9.0	20	13.0	34	7.8	20	8.1	30	12.9	22	9.7	28	9.0	28
LT	10.0	43	7.5	39	10.5	57	6.1	47	6.7	56	10.2	35	8.0	42	7.4	41
LU	11.3	66	8.5	72	12.5	83	7.6	79	7.8	104	12.0	55	10.0	64	8.7	68
LV	9.5	33	7.1	24	9.9	44	5.8	26	6.3	40	9.5	30	7.7	36	6.9	38
MD	13.9	23	10.4	18	14.6	31	7.3	20	9.1	27	11.4	24	10.6	26	10.2	25
ME	14.0	24	10.5	18	15.2	32	9.1	19	9.4	28	13.8	21	12.0	24	10.6	25
MK	10.6	28	7.9	20	11.8	35	6.0	22	7.2	29	9.0	30	8.8	30	8.2	30
MT	15.9	125	12.3	159	20.6	131	7.2	259	12.7	187	11.0	130	11.3	124	13.1	103
NL	11.1	173	8.3	226	11.7	196	7.3	256	7.4	285	11.4	153	9.6	178	8.2	201
PL	11.2	40	8.4	36	12.1	48	7.2	40	7.6	50	11.5	34	9.3	41	8.5	43
PT	9.8	54	7.5	52	12.4	64	5.1	70	7.7	71	7.6	56	7.9	53	8.6	45
RO	13.8	39	10.3	38	14.8	51	7.8	47	9.2	52	12.0	35	11.1	36	10.3	35
RS	13.7	24	10.1	19	14.7	32	7.4	21	9.1	28	11.2	25	11.0	25	10.2	26
SE	7.9	93	6.0	101	8.2	122	5.2	109	5.3	152	8.6	75	6.8	92	5.8	106
SI	13.8	62	10.3	71	14.2	75	8.1	93	8.8	104	12.6	53	11.1	57	9.9	60
SK	12.8	30	9.5	25	13.5	40	7.6	27	8.4	40	12.0	28	10.3	32	9.4	33
TR	9.9	38	7.5	30	11.4	46	5.4	38	7.0	45	8.4	37	8.0	38	7.8	37
UA	13.7	24	10.2	18	14.5	31	8.1	19	9.1	27	12.7	22	11.1	25	10.2	25
UK	10.2	72	7.6	78	10.5	92	6.9	88	6.6	115	10.7	60	8.9	70	7.4	80

TABLE 9.A.3 Average Yield (ton d.m./ha/year) and Roadside Cost (€/ton d.m.) High Management Level for EU28

	Miscanthus		Switchgrass		Giant Reed		RCG		Cardoon		Willow		Poplar		Eucalyptus	
Country	Yield	Cost	Yield	Cost	Yield	Cost	Yield	Cost	Yield	Cost	Yield	Cost	Yield	Cost	Yield	Cost
AL	18.2	858	13.6	1042	18.7	1712	10.6	2222	11.6	960	16.1	2115	15.0	1563	13.0	831
AT	12.1	141	9.0	154	12.7	222	7.6	240	8.0	181	12.0	215	10.1	228	8.8	131
BA	17.9	353	13.3	411	18.1	918	10.1	1105	11.3	349	15.0	1193	14.6	775	12.6	297
BE	12.0	63	8.9	70	12.2	80	7.6	76	7.7	103	11.6	60	10.0	68	8.5	78
BG	17.9	1297	13.4	1470	19.5	2068	9.1	2341	12.0	1333	13.8	2197	14.0	2480	13.6	1164
CY	20.4	4062	16.1	7033	26.9	7635	9.6	6733	16.7	6670	14.6	6224	14.8	7549	17.1	7054
CZ	11.8	36	8.8	33	12.8	130	7.8	125	8.0	91	12.3	124	9.9	148	8.9	75
DE	11.7	81	8.7	95	12.5	531	7.6	585	7.8	363	12.0	536	10.0	619	8.7	272
DK	10.2	113	7.7	133	10.7	140	6.4	152	6.8	179	10.8	93	8.5	115	7.5	122
EE	8.5	37	6.4	28	8.9	49	5.5	33	5.7	48	8.6	35	7.2	41	6.3	46
EL	20.2	3509	15.8	4885	25.1	5372	9.4	5744	15.3	4059	14.2	5380	14.8	6403	16.8	4033
ES	19.7	2757	15.1	3691	22.9	4293	10.0	4852	14.1	3273	15.0	4480	15.1	4990	15.5	3369
FI	7.2	100	5.4	111	7.4	172	4.6	164	4.8	192	7.7	117	6.1	141	5.2	129
FR	15.0	602	11.2	657	15.7	1250	8.8	1486	9.7	688	13.3	1469	12.2	1364	11.0	583
HR	16.5	702	12.5	670	18.2	1203	8.7	1552	11.2	545	13.0	1537	13.1	1383	12.7	473
HU	16.0	904	11.9	1005	17.4	1542	8.8	1841	10.8	893	13.3	1749	13.0	1820	12.2	753
IE	9.0	95	6.8	106	9.0	123	6.3	111	5.8	149	9.6	79	8.3	89	6.3	109
IT	17.5	1839	13.4	1981	20.0	2495	8.6	3486	12.3	1566	13.0	3265	13.1	3476	13.8	1603

KS	14.1	724	10.4	769	15.2	1286	9.1	1165	9.4	769	15.0	1159	11.4	1416	10.5	570
LT	10.0	43	7.5	39	10.5	57	6.1	47	6.7	56	10.2	35	8.0	42	7.4	41
LU	11.3	66	8.5	72	12.5	83	7.6	79	7.8	104	12.0	55	10.0	64	8.7	68
LV	9.5	33	7.1	24	9.9	44	5.8	26	6.3	40	9.5	30	7.7	36	6.9	38
MD	15.9	559	11.8	541	17.4	997	8.7	1048	10.8	485	13.6	1069	12.6	1157	12.1	424
ME	16.2	681	12.1	698	17.7	1233	10.6	1296	11.0	687	16.1	1379	14.1	1290	12.4	598
MK	18.0	1152	13.4	1297	19.3	1816	9.8	2204	11.8	1219	14.7	2066	14.4	2170	13.4	1090
MT	18.4	3664	14.3	5802	24.0	5858	8.5	6132	14.8	5038	12.8	5324	13.1	6443	15.3	4975
NL	11.1	178	8.3	226	11.7	196	7.3	256	7.4	285	11.4	153	9.6	178	8.2	201
PL	11.2	40	8.4	36	12.2	72	7.3	70	7.7	64	11.7	65	9.4	78	8.6	55
PT	19.5	2115	15.0	2122	22.8	2374	9.2	3977	14.1	1506	13.8	3813	14.3	4026	15.7	1898
RO	15.1	660	11.3	749	16.4	1194	8.5	1356	10.2	769	13.2	1309	12.3	1456	11.4	652
RS	16.4	989	12.2	1141	17.6	1703	8.9	1976	10.9	1023	13.4	1891	13.2	2009	12.3	898
SE	7.9	93	6.0	101	8.2	122	5.2	109	5.3	152	8.6	75	6.8	92	5.8	106
SI	13.9	62	10.3	71	14.6	75	8.4	93	9.1	104	13.0	53	11.5	57	10.2	60
SK	13.7	546	10.2	577	14.9	1080	8.4	1280	9.3	672	13.2	1256	11.3	1343	10.4	553
TR	19.3	2490	14.6	3096	21.6	3785	10.3	4432	13.2	2889	15.9	4059	15.1	4669	14.8	2573
UA	13.7	24	10.2	18	14.6	31	8.1	19	9.2	27	12.8	22	11.2	25	10.2	25
UK	10.2	72	7.6	78	10.6	206	6.9	179	6.7	177	10.8	116	9.0	193	7.4	131

Chapter 10

Research Overview, Gaps, and Recommendations

Calliope Panoutsou, Hans Langeveld, Uwe R. Fritsche, Berien Elbersen, Ilze Dzene, Rainer Janssen, Birger Kerckow, Dragoslava D. Stojiljkovic and Petar M. Gvero

10.1 INTRODUCTION

A sustainable and resource efficient development of the bioeconomy requires the establishment of continuous, secure, and cost-effective supply chains, that can ensure the provision of raw materials of known and consistent quality[1]. Most of the recent assessments and projections for biomass supply have been driven by the demand of policy and industrial actors in the bioenergy sector trying to learn how much biomass would be available at what price and what impacts of bioenergy development should be anticipated. As such, the key assumptions used for the estimation of available biomass quantities and the respective units in which they have been expressed are mostly related to energy. As the biobased economy evolves to cover a wider range of markets with variable end products, it is important that future research carefully examines synergies and potential conflicts and interdependencies among the different feedstocks and their applications to develop comprehensible ratios on composition and availability factors that support energy and GHG efficient uses of the biomass. The first attempt to harmonize assumptions and systemize a methodology for biomass assessments in Europe has been made during the period 2008–10 in the framework of the Biomass Energy Europe (BEE)[2] project. The project aimed to review methodologies for biomass assessments for energy purposes in Europe and its neighboring countries and to harmonize their assumptions in order to improve consistency, accuracy, and reliability as well as serve future planning towards a transition to renewable energy in the European Union. The major focus of the work has been on methodological and data harmonization for forestry,

1. Europabio: Building a Bio-based Economy for Europe in 2020.
2. http://www.eu-bee.com/.

Modeling and Optimization of Biomass Supply Chains.
DOI: http://dx.doi.org/10.1016/B978-0-12-812303-4.00010-0

energy crops, and residues from traditional agriculture and wastes. Following the European Bioeconomy Strategy[3], it is important that future assessments for potential supply should carefully examine synergies, conflicts, and interdependencies among feedstocks; they should therefore develop coherent indicators and coefficients accounting for coproduct allocation across value chains.

10.2 RESEARCH OVERVIEW

Current focus of European research for modeling biomass supply and logistics is to develop robust methodological frameworks with respective databases and inform a resource-efficient bioeconomy that will be resilient to climate change, together with a supply of raw materials, to meet the needs of a growing global population within the sustainable limits of the planet's natural resources.

Bioeconomy, including bioenergy and bio-based industries, is strongly coherent with the overall European goal of sustainable development because biomass resources when properly managed are renewable and therefore potentially sustainable. Use of bio-resources and conversion to energy and products can contribute to some extent to each of the six grand challenges.

While the importance of bioeconomy for bioenergy and bio-based materials has high priority in policy, industry, and research, there is still an intensifying debate about the extent of the potential impacts on food security, land use, and potential trade-offs among competing uses. These are likely to improve only with technological advance and innovation including clarity on issues regarding the potential biomass supply assessments and respective projections.

The structure of the European biomass supply industry is diverse in terms of feedstock types, availability, quality, and cost. Recent EU-funded projects, including Biomass Futures (www.biomassfutures.eu), Biomass Policies (www.biomasspolicies.eu), and S2Biom (www.s2biom.eu) that provided evidence for this book, have advanced scientific knowledge around these issues by improving methodologies and datasets for estimating biomass supply potentials across Europe. They have applied harmonized scientific approaches, provided guidelines for data collection, developed a set of availability and residue to main product ratios, and made this knowledge publicly available through reports and online toolsets.

Based on the sequence of research conducted during the last 10 years and presented in this book, recommendations for future research in the field are grouped in the following areas:

- Integrate sectors in land use and economic modeling to provide evidence for resource competition in biomass supply and logistics.

---

3. http://ec.europa.eu/research/bioeconomy/index.cfm?pg=policy&lib=strategy.

- Harmonize sustainability criteria and indicators across biomass feedstock types and ensure they are representative for the climate and ecology of the reference regions.
- Improve system analysis to inform policy at European, national, regional, and local level.

The following sections present detailed information for each of the three identified research areas in terms of:

- mapping past, and ongoing research projects and further group them to specific subtopics where appropriate;
- identifying gaps in research knowledge;
- listing a table with each priority and the respective timeline for research.

10.3 RESOURCE COMPETITION

10.3.1 Mapping Research Activities

Research funded so far in this area focused on assessments of biomass potentials at different implementation levels addressing various assumptions for displacement effects at different temporal resolutions (covering a projection period from 2000–30) (Table 10.1).

10.3.2 Knowledge Gaps

The main knowledge gaps for providing evidence-based information for resource competition can be summarized as follows:

- Harmonized and complete datasets
- Crop knowledge database on short rotation coppice (SRC)/short rotation forestry and perennial grasses, including yields, farming practices, and environmental impacts (greenhouse gas, GHG, emissions, water, biodiversity, etc.)
- Integrated land use modeling and impact assessments at European and national level considering all land use sectors and uses of biomass (food, feed, bioenergy, and bio-based products)
- Assessments of the most efficient use of biomass resources for the different competing uses.

New opportunities arising both from the INSPIRE Directive implementation as well as from the Copernicus programme and the new Sentinel Satellites for the assessment of the forestry and agricultural potential should be capitalized; concepts, and services need to be developed.

10.3.3 Recommendations

Table 10.2 provides a set of recommendations for future research priorities as presented in Fig. 10.1.

TABLE 10.1 Research Funded in Europe to Provide Evidence for Competition of Resources, Including Land and Water

Research Topic	Funded Project
Adaption to and mitigation of climate change	*BACCARA*—Assesses the impact of climate change on tree species assemblages, rates the risks to forest productivity loss, and provides recommendations; develops a decision-support system aimed at balancing costs and benefits for the establishment of new forest composition in order to anticipate and mitigate the potential detrimental effects of climate changes on forest productivity.
	(FP7-KBBE-2008–1–2–06: Forecasting forest diversity under the influence of climatic changes and the consequences for stability and productivity of forest ecosystems)
	DOFOCO: Do forests cool the Earth? Reconciling sustained productivity and minimum climate response with portfolios of contrasting forest management strategies. The overall goal of DOFOCO is to quantify and understand the role of forest management in mitigating climate change (ERC-SG-LS9)
	SmartSOIL—Sustainable farm management aimed at *reducing threats to soils under climate change*
	CATCH-C—Compatibility of *agricultural management practices* and types of farming in the EU *to enhance climate change mitigation and soil health*
	(Both projects under FP7-KBBE-2011–1–2–01: Sustainable management of agricultural soils in Europe for enhancing food and feed production and contributing to climate change mitigation)
	FP7-KBBE-2012: "Development of management strategies for planted and managed forests to increase mitigation capacity"
Biomass potential assessments	*BEE*—Biomass Energy Europe was initiated to *harmonize methodologies for biomass resource assessments* for energy purposes in Europe and its neighboring countries. The harmonization will improve consistency, accuracy, and reliability of biomass assessments for energy, which can serve the planning of a transition to renewable energy in the EU.
	(FP7-ENERGY-2007–3.7–01: Harmonization of biomass resource assessment)

	EUBIONET III aims to increase the use of biomass fuels in the EU by boosting sustainable and transparent biomass fuel trade, securing the most cost-efficient and value-adding use of biomass for energy and industry, and identifying yet unexploited biomass fuels. Among others: – 54 new types of biomass sources identified (Potentials 6.4–7.9) – New industry sectors found for increased biomass use – The appropriate use of biomass resources will be assessed by analyzing competition and price situation of woody biomass use in forest industry and energy sector. www.eubionet.net (IEE/07/777 IEE programme)
	Biomass futures assesses the role that biomass can play in meeting EU energy policy targets. It defines the key factors likely to influence biomass supply, demand, and uptake over the next 20 years (meeting the RED targets). Among other factors, partners examine the EU heat, electricity CHP, and transport markets; supply and demand dynamics; the effects of indirect land use change, water use, and social aspects on future biomass supply, etc. (IEE/08/653 IEE programme)
	CEUBIOM—Classification of European Biomass Potential for Bioenergy Using Terrestrial and Earth Observations
	The aim of the project was to develop a platform and a self-sustained e-service to directly assist and train professionals from the Earth Observation (EO), agricultural and EO/biomass sectors about the *new, common and harmonized applications of EO* and a better understanding of each other's requirements. (FP7-ENERGY-2007–3.7–01: Harmonization of biomass resource assessment)
	RECOVER—Science-based remote sensing services to support Reducing Emissions from Deforestation and Forest Degradation (REDD) initiative and sustainable forest management in tropical region. Its main research focus is to develop a sound statistical concept and accuracy assessment procedure that enables the generation of more reliable estimates for forest degradation and change, as well as enhanced biomass estimates (FP7-SPACE-2010–1)
	REDDAF—Reducing Emissions from Deforestation and Degradation in Africa. The services and products that will be delivered to the user community include *forest cover maps and forest cover change maps* for 1990–2000 and 2000–2009/10 (land use changes based on six IPCC compliant land use classes); degradation maps, biomass maps, and the relevant digital datasets (FP7-SPACE-2010–1)

TABLE 10.2 Research Priorities for Providing Evidence-Based Information to Competition for Resources

Term	Priority
Long	Resource assessment "cross-sector" forecasts/monitoring guidelines in relation to resource efficiency, cascading factors, competition for land use, and implementation level (top down and bottom up)
Long	Develop biomass cost–supply curves in relation to cascading factors, costs, and life cycle analysis
Short	Database and large datasets management
	Crop knowledge database including yields, farming practices, and agro-environmental impact (GHG emissions, water, biodiversity, etc.)
	Feedstock quality data (physical and chemical) both for dry and wet biomass in relation to diverse end-use options and postharvest operations such as size reduction, densification, blending, etc.
Short	Locate the "hot spots" of bioenergy and bio-based materials. Assess the regional balance of biomass potentials with respective demand. The idea would be to match biomass potentials with the best local solution.

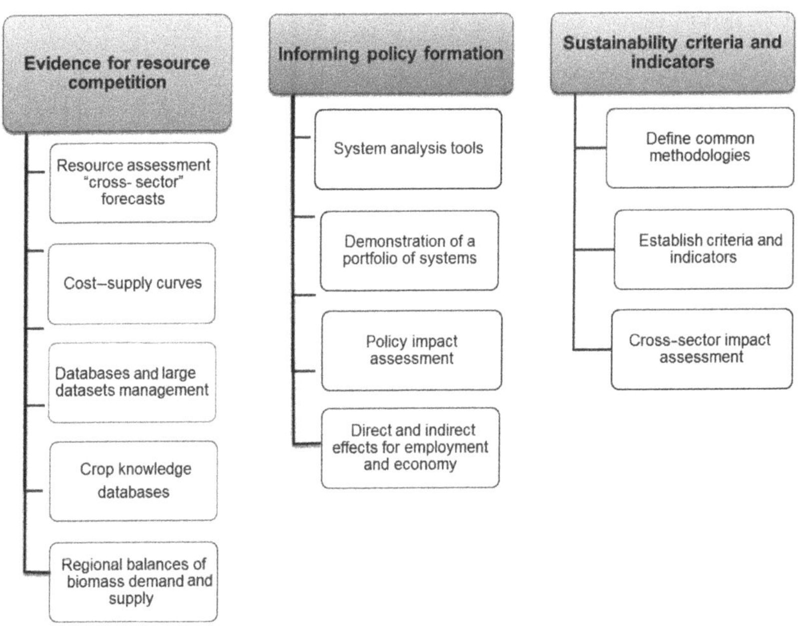

FIGURE 10.1 Future research in modeling biomass supply.

10.4 SUSTAINABILITY CRITERIA AND INDICATORS

The development of sustainability criteria with aligned indicators across feedstocks requires consistent efforts which should be aligned with the provisioning regional climate and ecology, rural development, and nature preservation priorities.

10.4.1 Mapping Research Activities

Sustainability for bioenergy (European Commission. SWD (2014) 259 final. State of play on the sustainability of solid and gaseous biomass used for electricity, heating and cooling in the EU. Brussels, 28.7.2014) has been a key issue in the formulation of the legally binding criteria of the Renewable Energy Directive (RED) and Fuel Quality Directive (FQD) since 2005, but the current EU legislation only addresses biofuels and liquid bioenergy carriers. (Directive 2009/28/EC of the European Parliament and of the Council of June 5, 2009, on the promotion of the use of energy from renewable sources and amending and subsequently repealing Directives 2001/77/EC and 2003/30/EC. The RED requires the EU to generate 20% of energy from renewable sources by 2020, and each Member State to achieve a 10% share of renewable energy sources in the transport sector. For ease of reference, it is recalled that the sustainability criteria for biofuels and bioliquids of the RED are as follows: Article 17(2) establishes minimum GHG saving values of 35%, rising to 50% on January 1, 2017, and to 60% from January 1, 2018, for biofuels and bioliquids produced in installations in which production started on or after January 1, 2017. According to Article 17 (1) wastes and residues only need to fulfill the minimum GHG requirements, not the other criteria. Articles 17(3), 17(4), and 17(5) require that raw material should not come from high biodiversity value areas, from the conversion of high-carbon stock areas, or from undrained peatland, respectively. Article 17(6) requires that agricultural raw materials cultivated in the Community are obtained in accordance with specific agricultural regulations of the EU. Article 18(1) requires that economic operators show compliance with the criteria using the "mass balance" method for verifying the chain of custody. Biofuels and bioliquids which do not meet the sustainability criteria cannot be counted toward the EU's renewable energy targets or the targets of the FQD (Directive 2009/30/EC) and national renewable energy obligations or benefit from financial support.) Since 2008, several communications from the Commission and EU-funded projects and studies (4FCrops, BioBench, BiomassFutures, BioTop, Crops2Industry, EEA, and Joint Research Centre, JRC, reports), as well as national (e.g., by Austria, Germany, Sweden, the Netherlands, United Kingdom) and international bodies (IEA, IEA Bioenergy, FAO, GBEP, UNEP, among others) broadened the debate to cover the sustainability of all bioenergy. Further work in the EU and beyond has been addressing the sustainability of the overall biomass use for nonfood purposes, i.e., including biomaterials, and biorefineries. As a part of that, significant improvement of knowledge on the sustainability

TABLE 10.3 Research Funded in Europe to Inform Sustainability

Research Topic	Funded Project
Addressing socioeconomic and environmental sustainability	*BioEnerGIS*—"GIS-based decision-support system aimed at a sustainable energetic exploitation of biomass at regional level." This project provides instruments to help decision makers to plan, on a regional scale, the sustainable energy exploitation of biomass. *A GIS-based decision-support system, named BIOPOLE, was developed to locate the most suitable sites—in terms of energy, environmental, social and economic sustainability—for biomass plants installations.* The aim was to find public and private stakeholders' interest in developing identified plants. Four regions, representing different environmental and economic zones, have been investigated: Lombardy (Italy), Northern Ireland (United Kingdom), Slovenia, and Wallonia (Belgium). (IEE/07/638 IEE programme-2007-BioBusiness)
	Global-Bio-Pact: (KBBE-2009–3–4–01) The objective of the Global-Bio-Pact project is the development and harmonization of global sustainability certification systems for biomass production, conversion systems, and trade in order to prevent negative socioeconomic impacts. Emphasis is placed on a detailed assessment of the socioeconomic impacts of raw material production and a variety of biomass conversion chains. The impact of biomass production on global and local food security and the links between environmental and socioeconomic impacts is analyzed. The interrelationship of global sustainability certification systems with international biomass trade of biomass and bioproducts and public perception of biomass production for industrial uses were studied while a set of socioeconomic sustainability criteria and indicators were developed, and recommendations were done how best to integrate socioeconomic sustainability criteria in European legislation and policies.

issues of forest bioenergy has been achieved, but also questions such as the carbon neutrality of forest bioenergy and biodiversity impacts of intensified extraction of agricultural and forestry residues are still controversial.

Building upon this debate on bioenergy sustainability, in November 2016 the Commission published the proposal on a revised Directive[4] on the promotion of the use of energy from renewable sources which was further revised[5] in February 2017. This revised Directive aims at strengthening the EU sustainability criteria by extending them to cover also solid biomass and biogas used in large heat and power plants above 20MW fuel capacity (Art. 26). However, other biomass uses for nonfood purposes such as biomaterials are still not covered in the revised Directive.

Recent research in the Biomass Policies and S2Biom projects focused on developing a harmonized approach with explicit criteria and indicators on how to "frame" the sustainability of the bioeconomy, in its environmental, economic, and social dimension for lignocellulosic biomass. It built on the existing knowledge available on the Member State and EU levels and integrated the work from the European Commission and JRC on the sustainability domain as well as the international domain (through IEA Bioenergy and GBEP). Systematic guidelines for harmonized methodologies to measure and assess respective impacts were developed (Table 10.3).

10.4.2 Knowledge Gaps

The main gaps in knowledge for informing policy formation and updates at European, national, regional, and local level can be summarized as follows:

- Common methodologies for sustainability analysis
- Life cycle analysis (LCA) of different biomass supply chains incorporating not only GHG, but also considering water use, biodiversity, land use change, and socioeconomic factors
- Criteria and indicators applied in cross-sector impact assessments.

10.4.3 Recommendations

Table 10.4 provides a set of recommendations for future research priorities as presented in Fig. 10.1.

10.5 POLICY FORMATION

Policy formation for biomass supply and logistics still requires consistent efforts which should be aligned with the domestic feedstock types, their sustainable management practices, rural development, sustainability, and nature preservation priorities (Table 10.5).

4. COM(2016) 767 final of 30.11.2016.
5. COM(2016) 767 final/2 of 23.02.2017.

TABLE 10.4 Research Priorities for Sustainability

Term	Priority
Short	Define common methodologies for sustainability assessment across feedstock types
Medium	Establish and validate a common set of criteria and indicators for sustainability in bioenergy and bio-based products
Long	Cross-sector impact assessments

Furthermore, the European Commission, DG Research & Innovation has also commissioned a study on "Research and Innovation perspective of the mid- and long-term Potential for Advanced Biofuels in Europe". The overall goal of the study has been to contribute to future policy developments in the area of sustainable low-ILUC biomass feedstock for advanced biofuels. Results of this study have informed the discussion on the role of research and innovation for biomass feedstock availability and sustainability for energy for the time horizons of 2030 and 2050.

10.5.1 Mapping Research Activities

Research funded so far in policy formation can be grouped in the following categories:

- Regional bioenergy planning and mobilizing biomass from agriculture and forestry in a resources-efficient way along the supply chain
- Socioeconomic and environmental sustainability of biomass supply and logistics
- Standardization and certification
- Promotion of market uptake for biomass, bioenergy, and bio-based products.
- Promoting bioenergy production on underutilised land.

10.5.2 Knowledge Gaps

The main gaps in knowledge for informing policy formation and updates at European, national, regional, and local level can be summarized as follows:

- Comparison of the efficiency of different existing support schemes for biomass feedstock mobilization at EU/national/regional/local levels
- Understanding motivations and decision making among farmers and forest owners in Europe as a basis for improved policy approaches
- Understanding of the economic, social, and environmental impacts of large-scale biomass feedstock systems using under-utilised lands for biomass feedstock production
- Intra-trade issues and imports.

TABLE 10.5 Research Funded in Europe to Inform Policy

Research Topic	Funded Project
Regional bioenergy planning and mobilizing biomass from agriculture and forestry in a resources-efficient way along the supply chain	*Biomass Trade Centres*—Project supporting the organization of regional markets for wood fuels, by creating Biomass Logistic & Trade Centres (BLTCs) and improving the professional approach of wood fuels producers, through training activities and demonstration events. The project developed guidelines on how to set up a regional BLTC and successfully BLTCs were set up during the project. (EIE/07/054 IEE programme)
	BiomassTradeCentreII—Follow-on of Biomass Trade Centres project, with the aim objectives of fostering the creation of biomass trade and logistic centers, but also putting effort in promoting the introduction of quality assurance and control systems. (IEE/10/115 IEE programme)
	Sucellog – *The project aims to widespread the participation of the agrarian sector in the sustainable supply of solid biofuels in Europe. Project action focuses in an almost unexploited logistic concept: the implementation of agroindustry logistic centres in the agroindustry as a complement to their usual activity evidencing the large synergy existing between the agroeconomy and the bioeconomy (IEE/13/638 IEE programme).* *Wood Heat Solutions*—The project aimed at *mobilizing the large biomass potential from unmanaged/under-managed forests for heat production in Croatia, Slovenia, and in the United Kingdom, by using the exemplar experience from Austria (on wood fuel quality and certification) and Finland (on the entrepreneurship model).* This was done through workshops, training courses, study tours to Austria and Finland, and expert support to specific projects. CEN standards for solid biofuels were also disseminated. (IEE/07/726 IEE programme)
	BEn—"Biomass Energy Register for sustainable site development for European Regions" supports *local communities' energy planning* through the development of a GIS *regional energy register* indicating local energy sinks as well as biomass potentials for energy production, in four European regions (North West/United Kingdom, Emscherippe/Germany, Pojezierze Gostyninsko-Wloclawskie/Poland, and Umbria/Italy). The project also aims at the creation of *regional biomass networks,* and the development of master plans for sustainable bioenergy planning including guidance for management and financing biomass energy investment, and at the implementation of bioenergy actions in the selected regions. (IEE/07/595 IEE programme-2007-Biobusiness)

(*Continued*)

TABLE 10.5 (Continued)

Research Topic	Funded Project
	The *MAKE-IT-BE* project focuses on the creation and implementation of *decision-making tools for local and regional policy makers in four EU regions* (in Austria, Italy, Slovenia, and United Kingdom). These tools are aimed at assisting policy makers in extensively identifying, evaluating, and initiating bioenergy chains. Thanks to the implementation of the tools the four EU regions have implemented Bioenergy Agendas with specific bioenergy projects. (IEE/07/722 IEE programme-2007-BioBusiness)
	Economic and ecologic assessment of agro-forestry systems in agricultural practice (Germany: four integrated projects: 22004907/22009707/22009807/22009907[i])
	4FCrops: Future crops for food, feed, fiber, and fuel
	The project surveyed and analyzed all the parameters that would play an important role in successful nonfood cropping systems in the agriculture of EU27 alongside the existing food crop systems. *The cropping possibilities based on regional potential levels, ecology, and climate were determined. A comparative cost analysis of nonfood and conventional crops was done. Socioeconomic impacts, like farmers income, rural development, public development, and public acceptance, was analyzed. Then environmental implications will be assessed compared to their respective conventional products* (fossil energy, conversional materials) (WP4). (KBBE-2007–3–1–07: FUTURE CROPS—technical, socioeconomic, environmental, and regulatory aspects of future nonfood crop systems)
Addressing socioeconomic and environmental sustainability of biomass feedstocks	*Crops2Industry*: Non-Food Crops-to-Industry schemes in EU27. The objectives of this project were to: *Explore the potential of nonfood crops*, which can be domestically grown in EU27 countries, for selected industrial applications, namely oils, fibers, resins, pharmaceuticals, and other specialty products*; Identify current molecular genetics technologies* (genomic and biotechnological tools) and suggest their potential applications in a crop-specific manner *to address a wide range of breeding constraints regarding yields and tolerance to abiotic and biotic conditions*; Explore the potential and feasibility of the European industry to make high-value bio-based

	products from renewable agriculture and forestry feedstocks and biotechnological routes; Perform *supply chain cost analysis, identify best business opportunities, and assess the socioeconomic impacts of selected crop-to-product schemes at EU27, regional and country levels*; Assess selected production and environmental impacts and identify a "core" list of standards and criteria for the environmental and socioeconomic sustainability of selected nonfood crops-to-industrial-products systems; Perform an overall assessment aiming to select and prioritize crops-to-products schemes in technical, socioeconomic, and environmental terms (FP7-KBBE-2008–3–1–03: European nonfood crops and their industrial application. Coordination and support action.)
	POPFULL—The objectives of the POPFULL project are: (1) to make a full balance of the most important GHG (CO_2, CH_4, N_2O, H_2O, and O_3), (2) to make a full energy and economic accounting; and (3) *to perform a full LCA of the global warming contribution of SRC*. The overall energy efficiency of the system will be assessed. (ERC-AG-LS9)
	ELOBIO assessed the impacts of growing demand for bioenergy on other markets (food, commodities), as well as the possible resulting ILUC effects. (EIE/07/139 IEE programme)
	BioEnergy Farm—The aim of this project is to increase the use and production of bioenergy and biofuels by farmers. This objective will be achieved by means of the creation of scan tools (both online and offline) that allow the assessment of the economic viability of bioenergy production in the farm (anaerobic digestion, energy crops, SRC, wood combustion). If the profitability and feasibility look good (both by a first online self-assessment tool and by an offline expert guided assessment tool), support for the implementation will be given, including the drafting of a business plan. This plan will help the farmer to obtain funds and to make an investment decision.
	(IEE programme)
	BioGrace aims *to harmonize calculations of biofuel GHG* emissions and thus supports the implementation of the EU RED (2009/28/EC) and the EU FQD (2009/30/EC) into national laws. (IEE programme)

(*Continued*)

TABLE 10.5 (Continued)

Research Topic	Funded Project
	PELLCERT—The key objective of the project *is to create and implement an ambitious and uniform certification system for pellets in Europe, called "ENplus," which will be used by both the heat and the power markets, for intra-European trade but also for imports*. This will be achieved through an intensive consultation with stakeholders. In addition, *procedures will be developed to certify the sustainability of pellet production and trade.* (IEE/10/463 IEE programme)
Standardization and certification issues	*SolidStandards*—The aim of the project is to support the implementation of EU standards for solid biofuels through an extensive training program in 11 EU countries and the implementation of the standards in selected companies. The feedback of solid biofuels industry players will be given back to CEN relevant committees for improving the existing standards and contributing to future standards. (IEE/10/218 IEE programme)
	BIONORM II—Prenormative research on solid biofuels for improved European standards. (FP6)
	EUBIONET III project aims to increase the use of biomass-based fuels in the EU by boosting sustainable and transparent biomass fuel trade, securing the most cost-efficient and value-adding use of biomass for energy and industry, and identifying yet unexploited biomass fuels. Among others: – Contribution to CN codes for wood pellets, to price index development for industrial wood pellets and wood chips and to CEN standards for solid biofuels. – Sustainability criteria for biomass fuels have been evaluated in cooperation with market actors. (IEE/07/777 IEE programme)
	SECTOR: Production of solid sustainable energy carriers from biomass by means of torrefaction
	Torrefaction and pelletization or briquetting are used to convert biomass into a high-energy-density commodity solid fuel or bioenergy carrier with superior properties for transport, handling and storage, and also in many major end-use applications (e.g., cofiring in pulverized-coal fired power plants, (co-)gasification in entrained-flow gasifiers and combustion in distributed pellet boilers or for biorefinery routes. The aim is to develop torrefaction and densification technology for a broad range of biomass feedstocks including woody biomass, forestry residues, agroresidues and imported biomass. Production recipes are optimized on the basis of extensive logistics and end-use testing. Experimental work is accompanied by desk studies to define major biomass-to-end-use value chains,

	design deployment strategies and scenarios, and conduct a full sustainability assessment. Results will be fed into CEN/ISO working groups and international sustainability forums. (FP7 ENERGY.2011.3.7-1: Development of new or improved sustainable bio-energy Carriers)
	BioBoost: Biomass-based energy intermediates boosting biofuel production The BioBoost project concentrates on *dry and wet residual biomass and wastes as feedstock for de-central conversion by fast pyrolysis, catalytic pyrolysis, and hydrothermal carbonization to the intermediate energy carriers' oil, coal, or slurry*. Based on straw the energy density increases from 2 to 20–31 GJ/m^3, enabling central scale gasification plants for biofuel production. *A logistic model for feedstock supply and connection of de-central with central conversion is set up and validated allowing the determination of costs, the number and location of de-central and central sites.* Techno/economic and environmental assessment of the value chain supports the optimization of products and processes. (FP7-ENERGY-2011–3–7–01: Development of new or improved sustainable bioenergy carriers)
Promoting a market for bioenergy carriers	Comparison of the efficiency of different existing public support schemes for biomass feedstock mobilization at EU/national/regional/local levels
	Understanding motivations and decision making among forest owners in Europe as a basis for improved policy approaches.
	FORBIO – The project develops a methodology to assess the sustainable bioenergy production potential on available "underutilized lands" in Europe (contaminated, abandoned, marginal, fallow land etc.) at local, site-specific level. This project will produce multiple feasibility studies in selected case study locations in three countries (Italy, Germany, and Ukraine) and thus demonstrate the viability of using land in EU Member States for sustainable bioenergy feedstock production that does not affect the supply of food and feed, in addition to not interfering with land currently used for recreational and/or conservation purposes. (Horizon 2020, No. 691846) SEEMLA – The main objective of the project SEEMLA ("Sustainable exploitation of biomass for bioenergy from marginal lands") is the establishment of suitable innovative land-use strategies for a sustainable production of plant-based energy on marginal lands while improving general ecosystem services. The use of marginal lands could contribute to the mitigation of the fast growing competition between traditional food production and production of renewable bioresources on arable lands. (Horizon 2020, No. 691874)

TABLE 10.6 Research Priorities to Inform Policy at European, National, Regional, and Local Level

Term	Priority
Short	Optimize supply system tools taking into account various land uses (including underutilised lands), resource efficiency, displacement effects, market interdependencies, etc.
Short	Demonstration of a portfolio of systems (subject to regional ecology and climate) with high potential for feedstock supply in relation to availability, infrastructure, and supportive policy framework.
Short medium	Supply and demand analysis and impacts for policy and financing mechanisms (local, regional level).
Short	Best practices, mobilization, benchmarking
Medium	Analysis and estimation of direct and indirect impacts over employment and economy at regional and national scale from specific value chains and the use/promotion of biofuels.

10.5.3 Recommendations

Table 10.6 provides a set of recommendations for future research priorities as presented in Fig. 10.1.

10.6 CONCLUSIONS

Biomass is an important domestic asset for European countries. It can be feedstock for energy and bio-based products and materials. It is widely available and well suited to a range of conversion routes and applications.

Nowadays, its market uptake is within traditional energy uses mostly while significant R&D is taking place to ensure compatibility with innovative conversion technologies and integration of supply and logistics with existing infrastructures in the forest, agriculture, and biowastes industries.

The future development of bioeconomy will require:

- alignment of the various research funding streams to ensure system integration for the research outputs and avoid overlapping;
- long-term consistency in R&D funding both in terms of budget and appropriately focused topics that will be oriented toward mobilizing and integrating the European feedstocks into day-to-day operations of the respective industries;
- monitoring past and current activities and building on these to ensure that European researchers, their work, and infrastructures are appropriately acknowledged and future research priorities capitalize on the knowledge generated so far.

Stimulation of biomass mobilization through markets and policy support without increasing the competition between food and nonfood biomass uses while at the same time boosting the biomass commodity market and ensuring increased security of supply for the biobased industries.

Based on the above, the suggested research priorities could be considered in the future by the various funding bodies and policy-making institutions.

FURTHER READING

Annevelink, B., Anttila, P., Väätäinen, K., Gabrielle, B., Galindo, D.G., Leduc, S., et al., 2017. Modeling biomass logistics.

Dees, M., Fitzgerald, J., Anttila, P., Glavonjic, B., Verkerk, H., Lindner, M., et al., 2017. Assessing lignocellulosic biomass potentials from forests and industry.

Elbersen, B., Forsell, N., Leduc, S., Staritsky, I., Witzke, P., Jacqueline Ramirez Almeyda 2017. Existing modeling platforms for biomass supply in Europe: inputs, outputs, and projection potentials.

Elbersen, W., Lammens, T., Alakangas, E., Annevelink, B., Harmsen, P., Elbersen, B., 2017. Lignocellulosic biomass quality: matching characteristics with biomass conversion requirements.

Londo, M., van Stralen, J., Mozaffarian, H., Uslu, A., Kraan, C., 2017. Modeling demand for bio-based sectors.

Panoutsou, C., Bauen, A., Elbersen, B., Dees, M., Stojadinovic, D., Glavonjic, B., et al., 2017a. Biomass supply assessments in Europe: research context and methodologies.

Panoutsou, C., Perakis, Ch., Elbersen, B., Zheliezna, T., Staritsky, I., 2017b. Assessing potentials for agricultural residues.

Ramirez, J., Monti, A., Elbersen, B., Staritsky, I., Panoutsou, C., Schrijver, R., et al., 2017. Assessing the potentials for nonfood crops.

Vis, M., 2017. Assessing the potential from bio-wastes and postconsumer wood.

Index

Note: Page numbers followed by "*f*," "*t*," and "*b*" refer to figures, tables, and boxes, respectively.

M

N

U

V

W

Z

CPI Antony Rowe
Chippenham, UK
2017-08-22 09:44